FORGOTTEN MONORAILS OF *Canada*

Paul Langan

ISBN -978-0-9685867-6-1

Forgotten Monorails of Canada

brentunespublishing@yahoo.ca

TABLE OF CONTENTS

INTRODUCTION

I am not exactly sure how my interest in monorails in Canada began. When I was young, I rode the monorail at Expo 1967 with my family. Much later in life I started taking an interest in stories about the enduring idea of a monorail in Niagara Falls, Ontario.

Then, in the early 1990's I became aware of Kim Pedersen and the Monorail Society that he started in the USA. His group was the main source of information on monorails.

About 10 years ago, I was looking at old photos of our family visiting Rainbow Valley Amusement Park in PEI. There were photos showing our family riding the novelty monorail around the park.

I asked myself: How many monorails were actually installed in Canada and how many were proposed over the years? The answers to these two questions form the basis of this booklet.

Although we have the Internet, Google and Wikipedia, this topic was obscure and involved a lot of old-fashioned research. Cold-calling people and going through over 70 years of old newspapers from each Canadian major city were important parts of my information gathering. At the back of this book, I have included a list of the newspapers that I used to piece together the history of monorails in Canada.

I want to thank my wife, Dr. Debra Langan, for her editorial work.

I also want to make a special mention of two people that helped in my research: Richard Goodell and Earle Davison.

Richard Goodell's dad, Muriel Goodell, was - in my opinion - the father of modern monorail in the USA. Often overshadowed by the Disney monorail, Muriel's monorail company (Monorail Inc.) had people riding their monorail in Texas in 1956. This was three years before the Disney Corporation started carrying passengers.

Muriel Goodell tried unsuccessfully to sell his company's monorail system in Edmonton, Niagara Falls, Toronto and Montreal. Richard corresponded with me regularly to help me understand the trials and tribulations his dad faced as an early supporter of monorails.

Earle Davison started up and ran the Rainbow Amusement Park, Cavendish, Prince Edward Island (1969-2005). Earle's friendly manner and first-hand account of the building of the monorail there was greatly appreciated.

Enjoy reading this one-of-a-kind piece of Canadian history.

Paul Langan,
Hespeler, Ontario, Canada

December 2nd, 2020

WHAT IS A MONORAIL?

Here is the definition from the Monorail Society.

A single rail serving as a track for passenger or freight vehicles. In most cases the rail is elevated, but monorails can also run at grade, below grade, or in subway tunnels. Vehicles are either suspended from or straddle a narrow guideway. Monorail vehicles are wider than the guideway that supports them.

There are two main types of monorails. The straddle beam and the suspended monorail.

Straddle Beam - As the name suggests, the vehicle straddles over the beam. It used rubber tires to travel along the beam. See photo below of a typical straddle beam monorail bogie. The Vancouver Expo 1986 was a good example of this type of monorail.

Vancouver Expo 86 straddle beam monorail
– credit Reinhold 3 Kocaurek com

Suspended Monorail - Again, as the name suggests the vehicle was under the beam. Here are two examples of this type of technology.

The Wuppertal Schwebebahn suspension monorail in Germany is the oldest suspension monorail in the world. It opened in 1901 and still operating. The bogie system is wrapped around the side and top of the beam and the vehicle is underneath.

In Chiba Japan, the Chiba Urban Monorail started up in 1988. It is the world's longest suspension monorail. It is different than the German monorail as the bogie wheels and electrical are all inside the beam of the hanging vehicle.

Chiba Urban suspended monorail: credit https://commons.wikimedia.org/w/index.php?curid=1700954

Almost all monorails have electric motors that connect to some sort of electrical third rail that provides power to the trains.

Are Maglevs a monorail?

Magnetic Levitation trains certainly can be classified as a monorail if their design fits the definition which some do. The big difference is the Maglev trains do not touch the beam they travel on, whereas traditional monorails do.

What are examples of transit systems in Canada that are NOT monorails?

The easiest way to tell if a vehicle is not a monorail is if the vehicle travels on a guideway that is wider than the vehicle. There are other names for this type of transit such as an automated people mover.

Here are examples in Canada of systems that are not monorails: Vancouver Skytrain, Toronto Pearson Airport Link Train, and the former people mover at the Toronto Zoo.

Now let us look at the interesting history of monorail systems and dreams in Canada.

BRITISH COLUMBIA WENNER-GREN MONORAIL PROPOSAL - 1957

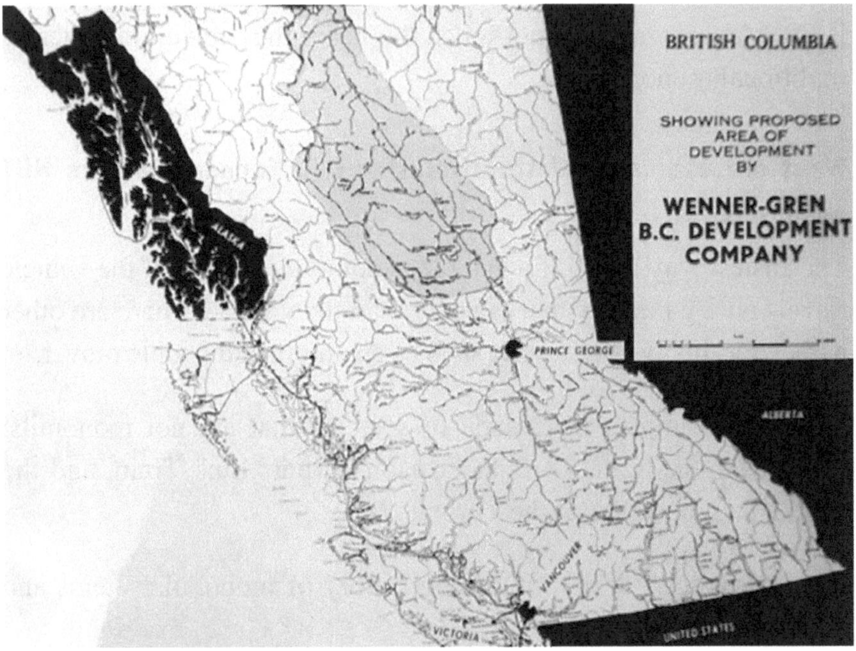

Wenner-Gren map. credit author unknown source swedish press

In February 1957, British Columbians were shocked by the news that Swedish Billionaire Dr. Axel Wenner-Gren had made a massive proposal for the development of northern B.C.'s Rocky Mountain trench.

Axel had invented the Alweg monorail in the 1950s also. It was a popular system that Disney and other systems around the world emulated.

The Wenner-Gren BC Development Corporation deposited a $500,000 guarantee with the BC government and then announced that they were prepared to spend $5 million exploring the possibilities of the 40,000 square mile Trench area.

In return for its $500,000 bond, the Wenner-Gren company received from the B.C. government a reserve on the lands and timber in the development area. The mineral reserve covers about 15% of the area.

The company would do surveys to see the potential wealth in the timber, minerals and possible hydro-electric power. If the survey shows justification for the development the company intended to undertake several projects.

Wenner-Gren committed to building a monorail that would travel at speeds of up to 289 km/h (180 mph). It would be 643 km (400 miles) in length running from Fort McLeod to Yukon border. It would move primarily freight (timber/minerals etc).

It would be constructed by Wenner-Gren who owned Alweg Company. They had a test track in Cologne, Germany that officials from the BC government visited in October 1952. Alweg monorail started to run in 1959 at Disneyworld, and in 1962 at the Seattle world's fair.

Just two days after the announcement of the project Vancouver Sun columnist Bill Fletcher on Feb. 14, 1957 was already asking some critical questions about the monorail proposal. He questioned the idea that a monorail would whip along the trench with cars loaded with lumber, ore or pulp at the 200mph speed. He asked some level-headed railroaders and they said it was ridiculous. The test track in Germany could hit 90 mph. Nobody suggested they could haul a payload of cars at anything like this speed. The principal reason being that the monorail was designed to carry passengers.

Premier Bennett was already asking the federal government on February 15, 1957 for a subsidy for the Wenner-Gren plan to be implemented.

Premier W.A.C. Bennett and Axel Wenner-Gren Victoria BC March 1957
Credit: Boman and Dhalberg Stockholm Sweden 1975

On February 19, 1957, Wenner-Gren announced there would be an interchange between the monorail and traditional passenger rail freight cars. The reality was that no such system existed between the two diverse forms of transportation.

On Feb.21,1957, Vancouver Sun columnist Bill Fletcher once again wrote a column that was scathing about the high-speed freight monorail. After talking to ALWEG monorail engineers in Germany, it became apparent that a high-speed freight monorail did not exist and there was no real solution to the interchangeability between the monorail and a typical freight train.

By March 11th, 1957, after persisting questioning from Bill Fletcher, Wenner-Gren conceded that if a traditional freight railroad was more practical than a monorail they would build. Wenner-Gren still preferred the monorail. Questions about the suitability/practicality of the monorail were increasing.

In October 1957 a trade mission from BC went to ride the ALWEG monorail at the test track and came back impressed with the ride.

In 1958 there was not a lot reported in the media except that surveys have suggested there was a great potential for strong financial returns on lumber/minerals and a hydro/electric project. The emphasis in 1958 seemed to shift dramatically to the hydro/electric prospect of the development.

On October 14th, the Vancouver Sun reported that the monorail line would have to be moved out of the Rocky Mountain Trench to avoid tremendous flooding planned there.

On September 7th, 1958, the Vancouver Sun newspaper reported that the Wenner-Gren B.C. Development Corporation had abandoned plans for a monorail through north central B.C. in favour of a conventional railway. The company received a report from Col. Sidney Bingham of New York who conducted the railroad survey for Wenner-Gren. He recommended they forget about the monorail and build a conventional railroad.

Although the whole history of the Wenner-Gren Rocky Mountain Trench and subsequent power plant projects are complicated. Suffice it to say the far-fetched high speed freight monorail scheme never saw the light of the day.

VANCOUVER - WELLS COATES PROPOSAL 1957

Wells Wintemute Coates was born on 17th December 1895 of Canadian parents in Tokyo, Japan.

He served in World War I, first as a gunner and later as a pilot with the Royal Air Force. He attended the University of British Columbia where he obtained a BA degree in May 1920 and a BSc degree in May 1922. In October 1922 he registered at East London College where he studied engineering, obtaining a PhD in 1924. During World War II, he again served with the RAF, this time working on fighter aircraft development,

He was an architect, designer, inventor and writer. He was, for most of his life, an expatriate Canadian who is best known for his work in England.

1957 Wells Coates Vancouver Proposal.
Wells Coates photo credit Wikipedia

As an architect, the three apartment buildings he designed in England are iconic examples of his work. The most stunning was the modernist block of flats known as the Isokon building (1934) in Hampstead, London. He also designed the Embassy Court in Brighton (1935) and 10 Palace Gate, Kensington (1939)

As a designer he was hired by E.K. Cole & Co. Ltd. (or 'Ekco') to develop a modern radio for the company's new bakelite material. Wells came up with a unique design. The AD-65 round shaped radio was available in black, walnut, ivory and even green. It is highly sought after by collectors.

Wells had a passion for sailing and boats. He designed a boat, called the Wingsail. It had a rigid sail design mounted on a catamaran hull. Unfortunately, he could find no company to produce it.

He returned to Vancouver for the last few years of his life. In what would be his last assignment he was to design a monorail rapid transit system for Vancouver, dubbed the Monospan Twin-Ride System (MTRS).

According to Laura Cohn (Wells daughter), in her 1999 book about her dad entitled, ``The Door to a Secret Room," the idea emerged from a research study he had undertaken in July 1957 for the British Columbia Electric Company in Vancouver. The purpose of this study was to assess changes in the design of mass rapid transit systems, "involving comparisons and evaluations and the development of new solutions in a form sufficient for economic analysis."

In fact, Wells had made some extensive notes three years earlier on a possible monorail system, in connection with the proposed new town of Iroquois for which he was a planning consultant. The BC

Electric contract gave him six months for their study and it was ready in October 1957 and produced as an "Interim Report'."

Dr. Wells Coates died of a heart attack while swimming off Vancouver beach on June 18, 1958 at the age of 63. His Monospan Twin-Ride System monorail never came to fruition.

VANCOUVER BILGER MONORAIL PROPOSAL 1972

Anson Bilger had been promoting monorails since the mid 1940s. Anson first came up in the media records when his wife received $100,000. in a divorce settlement (In today's dollars that would equal $1.7 million).

In 1946, he was pitching his monorail to the San Francisco's Board of Supervisors. In 1951 he had written an article for the US Department of Defense entitled: "Monorails - A Defense Weapon?"

In 1951 he also started franchising his monorails to companies. Surprisingly, Pioneer Machinery in Calgary Alberta was one the companies that was awarded a franchise to build industrial and passenger rail monorails. During this time, he even travelled to South Africa to try and sell his monorails.

Anson Bilger's monorail idea first showed up in Vancouver in 1970. A local firm, Columbia Rapid Transit Lined Ltd., wanted to sell the city council on a monorail line that would extend from Project 200 (a failed waterfront redevelopment project) to the airport. Columbia was the Canadian franchise for the Bilger Suspended Monorail System.

After the first proposal in 1970 failed, backers of the project came back in April 1972 to try and convince the City of Vancouver council again.

Bilger Monorail Concept – credit The Buzzer blog

Now Robert Craig, a Vancouver auto salesman, head of Colonial Transit Lines Ltd., stated his firm would build and operate the monorail if the city granted a right-of-way. Money to finance the project would come from US sources and the idea was that it could be completed in two years. Anson Bilger, developer of the suspended monorail, came to discuss his monorail which would be the one Colonial would be using.

The following information about the route and costs were obtained by viewing the 1972 preliminary report submitted by Bilger Monorail Company.

The Route

Starting at the Vancouver International Airport, the route might proceed northeasterly across Sea Island to a point in the general area of Richmond Island where a suspended crossing structure might easily be installed, thence connect with West Boulevard, thence along the Arbutus Street using the air rights along and above the existing railroad right-of-way as far as possible, thence some suitable route from West Broadway to Burrard to terminate at, near or even above the CPR yards at the foot of Burrard St.

The Trains

Anson never clearly outlined the schematics of what the monorail would entail. It was the suspended monorail. He did have patents on such things as the Monorail Wheel (patent 1949 US2488523A) and Monorail Brake (patent 1947 US2489740A).

A 3-car train - 300 riders maximum per trip in any direction. 6,000 riders for one train day. Assuming 10minute headway, this permits six such trains per hour. Thus 36,000 passengers can be handled by the trains daily.

Total Cost to build the Bilger Monorail was $35 million.

When the monorail presentation came before the civic and transit authorities on September 6th, 1972 there were definitely some that supported the Bilger monorail and some were against it. According to the Vancouver Sun article the following day, the most vocal on the negative side was Chairman Allan Kelly who told Bilger more than once, "You don't know what you're talking about". When Kelly's questioning revealed that Bilger does not have a working model of the monorail, Kelly stated, "I can see nothing at the moment to indicate we should take you seriously."

On the positive side, Mayor Dugald Morrison and Transport Committee Chairperson Bert Kennedy were strongly supportive of the monorail proposal and both unanimously condemned the attacks by Allan Kelly on Anson Bilger. In the Richmond Review, Sept. 8, 1972 Kennedy stated," Kelly is living in another century and I think it's the last century."

The Bilger monorail system was not built in Vancouver. Simple reasons for that could have been:

> They could not raise the $35 million needed to complete it.
>
> They had zero experience in actually building and operating a monorail system.
>
> They had no rights to use the CP Rail corridor.

Bilger Monorail ad. credit The Vancouver Sun June 26, 1971

VANCOUVER EXPO MONORAIL 1986

Vancouver Expo 86 monorail - credit Colin Arnot

At Expo 1986, the theme was "Transportation and Communication: World in Motion - World in Touch". It ran from May 2 until Monday, October 13, 1986 in Vancouver, BC.

Officials had hoped 13.7 million people would visit. By the end of the fair, more than 22 million attended, with a single-day record of 341,806 visitors (almost 80 per cent of Vancouver's population at the time).

The transportation theme at the Expo included a display of HSST (High Speed Surface Train) of Japan Air Lines. The magnet levitation train had a 450 meters demonstration rail. There were also gondola cars.

The 22-kilometer ALRT (Advanced Light Rail Transit) line, now known as the Skytrain, was completed in time to connect to the Expo site.

Amazingly, there were 39 major transit exhibits and demonstrations by 12 countries.

The $15-million monorail glided along the 5.6-kilometre route. Built by Von Roll/Habegger (Switzerland) It had seven stations and 10 trains, on the EXPO site. It transported up to 60,000 passengers a day.

Vancouver Expo 86 monorail. credit Colin Arnot

Alton Towers Theme Park

After the Expo closed in October 1986, the Von Roll Mark II Monorail was shipped to Alton Towers, Stoke-on-Trent, United Kingdom, where it was reconstructed through 1987 and ready for its grand reopening in August.

Here are the facts of the Alton Towers Monorail System.

> Von Roll Holding AG, Allot & Lomax, C.W.A Type: Monorail Automated People Mover
>
> Location: Towers Street; Cost: £15 million; Track Length: 2800 metres; Track Height: 9 metres
>
> Opened: 13 August 1987

The ride gained the name "the Silver Ride Monorail" and each of the trains was named after a planet. William Shatner, the original Captain Kirk, participated in the opening of the ride, even taking the opportunity to captain his own Monorail train. Shortly after the opening ceremony he was also joined by George Takei, who missed the ribbon cut due to a navigation error.

The monorail is still running at the park. If you ever have a chance to visit there, take a ride on what was once Canada's only full-scale monorail.

Expo 86 Monorails at Alton Towers during construction. credit towersstreet.com

CALGARY BRENNAN MONORAIL EXHIBITION - 1911

The Calgary Exhibition and Stampede was called the Calgary Industrial Exhibition from 1911-1918. An advertisement for the 1911 exhibition mentioned a special feature, Brennan's Monorail Car.

1911 ad. credit Alan Zakrison

The Brennan Gyro-Monorail (patented 1903) was developed by the Irish-born Australian inventor Louis Brennan. (1852-1932). It was 40 feet long and weighed 22 tons, and was designed to carry 10 tons. Speed on the level was 22 mph.

The vehicle was balanced by two vertical gyroscopes mounted side by side, and spinning in opposite directions at 3,000 rpm on a single rail. Each gyroscope was 3.5 feet in diameter and each weighed 3/4 of a ton. They were enclosed in evacuated casings to reduce air-friction losses. The rotational axes were horizontal.

In 1909 and 1910 demonstrations of the Brennan monorail with passengers occurred in London.

Brennan Gyro Monorail Demonstration Model. credit Stephen Holland

Louis also developed two working models. The first demonstration model was just 2 feet 6 inches by 12 inches (762 mm by 300 mm). Brennan produced a larger model, 6 feet (1.83m) long by 1 feet 6inches (0.46m) wide, kept in balance by two 5- inch (127 mm) diameter gyroscope rotors. That second model is still in existence and stored in a museum in London.

It is strongly suspected that it was one of the models that was brought over to be used at the 1911 Calgary Industrial Exhibition. Unfortunately, newspaper records from that time offer scant information on the Brennan monorail car at the exhibition.

Interesting to note that in 1910 there was some interest from Alaska about possibly using the Brennan monorail car in the mining industry. This never did happen.

The Brennan monorail failed to attract further investment. Of the two vehicles built, one was sold as scrap, and the other was used as a park shelter until 1930.

CALGARY (OGDEN) MONORAIL 1911-1917

In 1910 W. Maloney filed two patents for monorail conveying cars. In 1912 the Alberta Mono-Rail Company paid $200,000 to purchase six patents and caveats from the Alberta Mono-Rail Syndicate, including W. Maloney's two 1910 patents. The Syndicate held the rights in Alberta of the Canadian Mono-Rail Car Corporation of Montreal. AMRC's directors were J.M. Healy, H.H. Morency, G.A. Crooks, Charles Knight, Clifford B. Reilly, and George L. Peet, all of Calgary.

Monorail Engine And Cars

Photograph taken by The Herald Staff Photographer

2012 Alberta Monorail Co Engine and Car. credit The Herald

In 1911 at the Calgary Industrial Exhibition the Alberta monorail Co. set up a short temporary track. It was an outstanding attraction when rides in the gravel cars were offered.

The Alberta Monorail Co was incorporated in 1912. It built a monorail that ran from the Bow River to the J.F. Burns Sand & Gravel Company pit near Calgary. (Ogden area) The purpose was to haul gravel.

A principal shareholder in the Alberta Monorail Co was Walter Moss, who had subdivided a 1/4 section he called Maharg, known today as Riverbend. His land contained gravel deposits, and the company planned to build a track along the river to the Bonnybrook Bridge area. A good portion of the track was actually built and can be seen on a 1924 air-photo of the area. By late 1913, the line was in operation.

Photo showing the line - credit Glenbow Archives

Its monorail locomotive had been designed by Bernard Hervey, an Oxford-educated engineer with extensive CPR experience. Later in 1912, Hervey would become Superintendent of Dominion Parks for Canada.

The company also planned to use monorails as "feeders" to conventional railways, of which Alberta had many.

In early 1914, the property boom was fading fast, and with it the dreams of the Alberta Monorail Co shareholders. The company had planned to order 30-seat passenger cars. By 1917, the company had dissolved.

CALGARY TRANSYT MONORAIL PROPOSAL - 1982

From the Calgary Herald May 28, 1982. The Toronto company TranSyt President, Lyman Richardson, came to Calgary AB looking to build a 1.6 m test track for their monorail.

Backed by Saudi Arabian money, the company spent 18 months developing the monorail. TranSyt committed to building the test track for free if council accepted an offer from them.

TranSyt also offered the city an interest-free loan to add about 14.5km of monorail track to the experimental section, but Mayor Ralph Klein was only interested if the provincial government would fit the bill. TranSyt would take a letter to the city proposing the test track and potential 14.5 km track if the test track is successful. A line to the airport was one possible route.

1982 TranSyt Model. credit Calgary Herald

Mr.Richardson stated the test track could be built within two years if the city would provide the right of way.

TranSyt also had a small-scale version of the monorail to show to the media and politicians. Forward thinking Calgary Mayor Ralph Klein said he would support this idea at council. He was even willing to put the north-east line of the Light Rail Transit on hold for the monorail, if the province would pay for the system.

Confident the system would pay for itself; Richardson said the city could pay back the loan for the proposed 14.5 km of track out of revenues from the monorail, which could carry cargo as well as people.

Each monorail car running on tracks 10m above ground would carry 200 people. The system would cost about $9.5m a kilometre to construct. Richardson stated to the media that the monorail could reach a speed of 240km/h outside the city and about 120km/h in the city.

Transportation Commissioner Bill Kuyt said the monorail concept was interesting but the city should not get into the business of building an "unproven prototype" system.

The Calgary Herald October 22, 1982 reported that construction of a 5km test track for the first Canadian-built monorail could start as early as the following spring near the Calgary International Airport. TranSyt officials were going to meet with Transport Canada and city officials for the installation of the $20-million testing system northeast of the airport. It was repeated that the Canadian company was representing Saudia Arabian interests who would pay the full costs of the test track.

Detailed engineering plans were studied by Transport Canada officials in Calgary to ensure the oval test track would not interfere with airport operations.

The tone of the project had changed by March 21, 1983 when in a Calgary Herald article, TranSyt gave Transport Canada 30 days to support the Calgary monorail project or they might have moved the project to Philadelphia. Lyman Richardson was annoyed with the red tape and delays dealing with Transport Canada. He stated, "We're sweating blood to get this thing going and we can't even get them to say no."

On March 23rd, 1983, the City of Calgary strongly endorsed the monorail proposal. Mayor Klein had written a letter to the Minister of Transport Jean Luc Lepin to get involved with the project to smooth the way for the construction of the track. A Minister's assistant stated that a decision could be made in weeks after meeting with Minister Lepin.

By April 14, 1983, TranSyt officials were looking at a new site for their test track in Calgary. Delays in approval by Transport Canada had led TranSyt asking to build the 5km test track on a road allowance beside Deerfoot Trail between 64th and 112th Avenue N.E.

Mayor Klein was supportive of the test monorail but had to check with city officials to see if there were any problems with the new location.

Mayor Klein stated on December 3rd, 1983 that the City would not negotiate to buy an experimental monorail if it were to be built. The City has made no agreement with TranSyt Canada Inc. to buy the line if it proved successful.

Klein stated to the Calgary Herald, "The city would have the right of first refusal. We are not willing to negotiate. We made no commitment to buy it".

However, TranSyt president Lyman Richardson said a condition of the experiment was that the city would purchase the line if it were a success. He further stated it was a typical Canadian deal in that it goes on and on and nothing happens.

On May 24th, 1984, the Calgary Herald headline stated, "City Loses Test Track". TranSyt formally cancelled the project in Calgary. For all the reasons previously stated the project did not move forward.

Monorail Test Track Ad - credit Calgary Herald

CALGARY OTHER MONORAIL PROPOSALS

Nov. 1st, 1951 - Anson Bilger Calgary Franchise - Anson Bilger, President of General Monorail company announced the granting of an exclusive franchise to Pioneer Machinery of Calgary, for the province of Alberta, for construction and operation of both industrial and passenger monorail lines in that area. Mr Bilger did have several patents for component elements peculiar to balanced monorail. Although a big promoter of monorails Anson Bilger never built an operational system. There is no evidence to suggest Pioneer Machinery of Calgary did any better. (Daily Independent Journal - San Rafael California, Thu. Nov. 1, 1951, Page 17).

Mar. 1st, 1963 - Calgary to Banff Olympic Monorail - A proposal was made for a monorail tram to be built from Calgary to Banff and from Banff to Norquay. The idea was hopeful that Banff would be chosen to host the 1968 Winter Olympics. The cost would run into millions of dollars. The proposal was made by Jack Anderson of the Banff Advisors Council. (Edmonton Journal - Mar. 1st, 1963, Page 10).

RENDERING OF SAFEGE AERIAL MONORAIL AT ELVEDEN HOUSE

The City of Calgary Archives, Board of Commissioners Publications, Transit for Calgary's Future, page 46.

December 1967 - Calgary Future Transit Study - The City of Calgary hired Philadelphia Pennsylvania firm, Simpson and Curtin, to do a study on transit options for the city. The final report was titled, "Transit for Calgary's Future" for the Calgary Transit System. The recommended system was a high-speed transit in corridors of maximum potential. In the downtown core an elevated monorail or subway system were recommended. The study stated that a subway system would cost $15.5 million more than an elevated system (City of Calgary Archives, Prepared for the Calgary Transit System - 1967 Transit for Calgary's Future - Simpson and Curtin, Philadelphia, PA. Pages 46-49)

December 1984 - Sarcee Trail Planning Study - A $419,000 Sarcee Trail Planning Study was done.(Tsuut'ina Nation paid 20%, the City paid 30% and the Province paid 50% of the cost) The study was primarily seen as a tool to aid the Nation in deciding whether to agree to allow the road through the reserve. This was to be achieved by assessing the impacts the road would have, and by exploring the potential for development of reserve land for the economic benefit of the Nation. One of the Development Scenarios (#6- Super Shopping Centre and 'Disneyland-like Complex) suggested a monorail system.

It was characterized in the report as "the most imaginative and daring of the six proposed (scenarios). The entire 300acre family entertainment complex site was proposed to house a "super regional shopping centre and entertainment complex", akin to West Edmonton Mall and anchored by four department stores. It would include not only shopping, but also a range of indoor and outdoor activities. This "small Disneyland-like complex" was envisioned with a monorail system with stations that would link the different shopping areas within the mall with outdoor attractions and parking (December 1984 Sarcee Trail Planning Study, Main Report for the City of Calgary, Province of Alberta, Sarcee Indian Band).

EDMONTON - HOUSTON MONORAIL CORPORATION PROPOSAL - 1962

1961 Goodell Monorail. credit author's collection

Muriel Goodell, President of Houston Monorail and inventor of the Houston Trailblazer Monorail proposed in early 1962 that a monorail be part of the transit system in Edmonton. Excitement for monorails was high due to the operational monorail in Seattle for the World's Fair there. Goodell's monorail system was now operational in Dallas.

On March 17th officials from Houston Monorail visited Edmonton with their concept for the city. There had not been any proposals for the location of the monorail within the city. The proposal never went any further.

In 1964 another monorail proposal from a Canadian company, Taylor Woodrow, met the same fate.

EDMONTON - REGINA - WINNIPEG - ROYAL AMERICAN MIDWAY - 1966

1966 Royal American Monorail - credit Leader Post Regina

This probably could never happen again. The Royal American Midway introduced in 1966 a new ride - a monorail. They actually built the footings for the track in each city they were in. The midway travelled by train. Royal American controlled the Western "A" circuit which included Calgary, Edmonton, Regina and Winnipeg. The midway started in 1934 in Canada and ended in 1975. At its peak they needed 96 train cars to move the midway.

On July 26th, 1966 the Regina Leader-Post ran an article outlining just how involved it was to make this new monorail ride a reality in each city it was in. Preparation for the installation would start two months prior to the midway coming to town.

The monorail's owner and manufacturer was Carl Miler of Portland Oregon. About a month before the midway came to town, a crew came to install concrete footings so the monorail could sit exactly true. The advance work, which is necessary to speed the setting-up job, was unprecedented for a carnival. It was customary for a midway to be laid out 24 hours in advance of the show' arrival.

The monorail with a minimum height of 15 feet, had two-seater cars, to carry 4 adults or 6 children. The monorail had to be moved around the Canadian "A" Circuit on trucks. The ride was well accepted. Documentation exists which shows the Royal American Midway monorail was set up in Winnipeg, Regina and Edmonton in 1966.

EDMONTON ALBERTA MONORAIL LTD FORT SASKATCHEWAN TEST TRACK AND PROPOSAL 1980

Perhaps the most interesting monorail proposal occurred when Gerry Baptist, who was the owner of several trucking firms, started up a Fort Saskatchewan based company named Alberta Monorail (1980) Ltd. Gerry believed that monorails could be a solution to key urban transportation programs.

George Haw Alberta Monorail 1982- credit Edmonton Journal

According to a November 19,1982 Edmonton Journal article, Gerry invested $270,000. of his own money in the development of a prototype monorail system. Amazingly he built a 213 metre test track in Fort Saskatchewan and developed a prototype rail car. The monorail car that was tested was about 8m by 3.7m, with a potential seating capacity of 50 for the full-scale vehicle. The tests were conducted in the winter and summer of 1982. They had made some modifications and were at the point where detailed engineering could begin.

George Haw was the company manager in charge of the demonstration track/vehicle.

They estimated a two-car configuration would likely be used for city operation, with the vehicles averaging speeds of 70-80 km per hour. Gerry stated that another advantage of the proposed system was that virtually all the components could be made by companies located in Edmonton.

Bob David, city engineer and general manager of the transportation management division, said because of the city's major financial commitment to Light Rail Transit, it was unlikely monorail systems could have widespread application there.

Interestingly, both TranSyt Canada (based out of Toronto) and Alberta Monorail Ltd. (based out of Fort Saskatchewan) were both pitching ideas at around the same time. TranSyt was selling their system in Calgary and Alberta Monorail was in Edmonton.

On May 28th, 1983, The Edmonton Journal article on Alberta Monorail stated that Gerry Baptist had died. The firm was extremely interested in seeing monorail developed in the province, but lacked the capital to promote the idea much further without some industry or government backing.

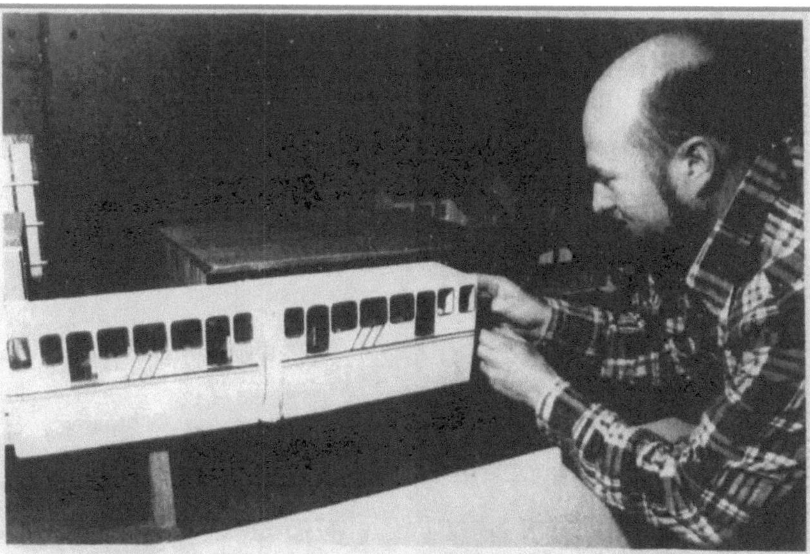

George Haw of Alberta Monorail with model of proposed skybound system
... the small firm hopes for government or industry backing

George Haw 1983 model monorail - credit Edmonton Journal

Unfortunately, this system was not built and the story of what happened to the Fort Saskatchewan test track and vehicle remains unsolved.

WINNIPEG MONORAIL PROPOSAL 1962

1962 Proposal. credit Winnipeg Tribune 1962-01-11, University of Manitoba Archives & Special Collections, Winnipeg Tribune, Collection 18-6

A Winnipeg Tribune article, January 1st, 1962, stated Metro Transit Director D.L. MacDonald would be visiting Seattle in the summer to see the World's Fair monorail.

Although the council had not discussed it yet, he felt they would be foolish not to look at monorail. MacDonald also felt they could afford a monorail system. He wanted it built down the middle of Portage Ave. and Main Street and it would cost about $4 million a mile. That would be cheaper, he said, than the $15-20 million for a subway.

In 1959 a transit report by Norman D. Wilson had suggested building a subway instead of expressways. The subways were never built.

MacDonald knew that a monorail would not have the capacity of a subway but he believed Winnipeg was not big enough to support a subway system. Macdonald stated," There is a great deal of fascination about the monorail. Basically, no city in the world has been sufficiently convinced about the virtues of monorail...before Toronto and Montreal built their subways, they both looked at monorails." He also wondered if the monorail could handle the Winnipeg weather.

Certainly, the biggest supporter for the monorail in 1962 was not evenly sure about choosing monorails. But he felt there was a real need for a modern transit system in 1962, and monorail was an affordable way to go. Sadly, his idea was never implemented.

WINNIPEG STEPHEN JUBA
MONORAIL PROPOSAL 1972 - 1977

On March 20th, 1972, the story broke in the Winnipeg Free Press with the headline, "Monorail Transit for the City was going to be studied." Mayor Steve Juba was examining proposals for a $22 million overhead rapid transit system.

Winnipeg Mayor Steve Juba Proposal.
Photo Mayor Steve Juba. credit Winnipeg Free Press

If built, the monorail system would probably run six or seven miles out Portage Ave from the corner of Portage and Main Street. A north-south branch would serve St. Vital and North Winnipeg, perhaps crossing the Red River into East Kildonan.

Deputy Mayor J. Paul Marion called the press conference after reports of the monorail project were aired on a local radio station. Mayor Juba was away on holiday.

The news article stated that the firm that would build the transit system was conducting a feasibility study on installing it for Winnipeg. City transportation officials had not been consulted nor had any of the commissioners or provincial officials.

Marion had suggested that 85% of the production work on the monorail system could be done in Winnipeg. The system would include two rails and would be held aloft, one going each way. Stations could be located in existing buildings. The trains would be electrically powered and would run on rubber wheels. The system would be quiet. Marion knew the firm was European but did not have the name of the company preparing the report for the monorail system in Winnipeg.

Councillor Marion supported the monorail idea even though it went against the 1969 city- funded study which suggested that Winnipeg should build a substantially more expensive subway.

Councillor Bernie Wolfe, chairman of the civil-transit authority, was critical of the monorail idea and wondered why there was no consultation by the Mayor on his monorail plan.

On March 27th, 1972 the Winnipeg Free Press came out with a scathing editorial against the monorail system titled, "Pipe Dreams." It was clear that - from the onset - the media and some councillors were dead set against the idea.

In the Winnipeg Free Press on May 9th, Mayor Steve Juba was upset with the newspapers, particularly the above- mentioned editorial for the negative stance on the monorail idea. He stated, "The critics hadn't based their opinions on the merits of the questions." He further chided the Free Press for not listing the author of the Pipe Dreams article. "They haven't got the darn guts to put their names to it.... They know who I am."

The Mayor had stated to the media that he was not proceeding with the scheme because of editorial opposition to it, claiming the two city newspapers had spoiled the political atmosphere.

On May 17th, the Free Press learned that the monorail proposal was prepared by Habegger Industries of Vancouver. Mayor Juba had been talking to the company about it since December 1971.Habegger Industries President, Kurt Hohenwarter stated the monorail would cost $1 million a mile and the foundations would cost an additional $75,000 to $150,000 a mile and each station would cost between $50,000 and $100,000. Habegger was known for their mini rail systems. The same system was used at the Expo in Montreal in 1967 and is still in use today.

The system would travel at a medium speed of 40mph (64kph). It would be a new model unlike the slower mini-rails they previously made.

The Mayor, at their Wednesday May 17th meeting, asked for a vote of confidence from city council on his monorail proposal but it was quickly sent to the executive committee without discussion. The mayor had asked in a letter to the councillors to decide whether they would consider it advantageous to meet with the monorail manufacturers.

In a May 27th article in the Winnipeg Free Press, Bob Lisoway stated that a lot of the controversy could have been avoided had Mayor Juba not been so secretive with details on the monorail project.

On July 26th, 1972, 40 city administrators/councillors, met for a full day seminar on Winnipeg's transportation problems. A film presentation on Mayor Steve Juba's proposed monorail was presented by two representatives of a Vancouver firm, Habegger Industries Ltd.

Councillor Bernie Wolfe once again attacked the plan stating that, "Winnipeg should not be a guinea pig in building an unproven urban transit system."

On February 23rd, 1973, another monorail manufacturer, Electronik Monorail Sales Ltd of Montreal made a presentation to the city's environment committee. That company had the Canadian franchise rights for the Hitachi-Alweg monorail systems which were straddle beam monorails. The Seattle and Disney World were two examples of their systems in operation.

On October 23rd, 1973 Mayor Steve Juba had a new idea to get his monorail system implemented. His plan called for the existing airport terminal to stay where it was, but the runways will be moved 10 miles (16km) northwest of the present location. The runways would be connected to the terminal by a rapid transit monorail system. The plan had been submitted to the federal government. Although persistent monorail critic Coun. Bernie Wolfe did not like the method of unveiling the plan, he felt it had some merit.

The plan was never approved by the federal government.

The Province of Manitoba at that time wanted to help urban centres with transit infrastructure projects. In March 22, 1974, the provincial Minister of Finance Mr Cherniack was asked if the province would support Mayor Juba's monorail plan. He stated they would not support any plan that had not been endorsed by council.

During the municipal election campaign in October 1974, Stephen Juba once again pushed the idea to the public and the media about a rapid transit modern monorail system that he stated was needed for Winnipeg. Juba won the election with 88% percent of the vote.

On June 27th, 1975, the monorail scheme was pulled out of the moth-balls when the civic executive policy committee recommended that the feasibility of building a rapid transit system be looked at again. A joint federal/provincial/municipal rapid transit study was being conducted.

December 3rd, 1975 - The preliminary results of the southwest transit corridor study suggested that light rail, bus lanes, or monorails could still be considered. Subways were ruled out as too expensive. Consultation with the public would continue until August 1976.

On August 6th, 1976 the Winnipeg Free Press reported that Mayor Stephen Juba came up with an idea that the future monorail rapid transit system could include luxury first class cars serving food, drink and liquor!

On December 23rd, 1976 Mayor Juba was upset, the transit study recommended diesel powered buses be used for the transit corridor. Juba stated, "They didn't even look at monorail. They have as much chance of implementing this report as a snowball in hell.

An executive committee report on January 7th, 1977, recommended electrically powered trolleybuses if the province paid 75% of the cost. Diesel powered buses were still the cheaper alternative.

In the spring of 1977 city council invited Habegger to submit a full proposal for a monorail. In August 1977 Habegger presented their Fixed Guideway System proposal to the city. Although the presentation response was generally positive recurring concerns regarding issues like whether the monorail could work effectively in Winnipeg's harsh winters came up.

The 1977 municipal elections featured Mayor Juba running again, 21 years after he first held the position. His primary opponents Robert Steen and Bill Norrie were not supporting the monorail system development. On October 6th, the Winnipeg Free Press featured profiles on all mayoralty candidates. On October 7th the headline of the newspaper stated, "Juba quits mayor's chair after 21 years'."

The main proponent of the Winnipeg monorail system was now out of politics. The monorail plan for Winnipeg became a memory.

ONTARIO SOUTHERN RAILWAY
MONORAIL - RIDGEWAY ON - 1896 - 1898

Ontario Southern Railway station. credit public domain

The Ontario Southern Railway was a short passenger-carrying monorail line which shuttled patrons of the amusement park at Crystal Beach, Ontario to and from Ridgeway, Ontario. It was in operation 1896 to 1898.

It was also known as the Ontario Southern Railroad, Peg-Leg Railway, and the Wonderful Bicycle Railroad.

Who owned the Ontario Southern Railway? The Times Review of Fort Erie, Jubilee Issue, July 18th, 1948 stated, "Captain Beecher,

Batavia N.Y. was manager of the railway. Other sources mention "Henry Beecher of Albion" as the designer and promoter. After two years, control of the road passed to Eber Cutier, who tried to reimburse himself for the wood he provided including the oak peg legs.

Eber Cutier of Ridgeway was the owner of a saw mill, dry goods store, and a builder. He built the Ontario Heritage designated home that is located at 348 Ridge Road N Ridgeway.

Where did the line run from? The northern terminus was a station near Young Lumber (approximately corner of Disher and South Mill St. Ridgeway, ON), a convenient connection for commuting Grand Trunk passengers. The Crystal Beach terminus was near the intersection of Erie and Derby Road. The line was approximately 2.4-kilometres long. To be exact the line was 1 1/2 miles,19 yds,1ft, 6 inches long.

Canada Southern Railway Line. credit Fort Erie Public Library

It was an early example of a monorail, consisting of a single running rail which was laid on top of a wood beam with guide rails mounted on each side. The majority of the line was elevated 3 to 9 meters, a single row of wooden legs (2.4m by 2.4m pillars of oak) thus giving rise to the nickname "the Peg-Leg Railway".

The line also passed through one small tunnel under a road, just south of Ridgeway.

It was said by older residents that at low speeds the little wheels (whether on one or both sides) were normally engaged. However, at the maximum speeds, estimated to be at 40 km/h, the monorail hurled over its ribbon of steel.

The passenger carriages were open sided and had a total of six benches in a back-to-back configuration. The cars measured1.5m by 4.5m, set upon the monorail track on two tandem wheels. A car had seats for 20 passengers. There were 2 carriages, only 1 being regularly used.

The "Wonderful Bicycle Railroad" had six wheels: two small wheels on each side projected under and within an angle-iron guide rail. This had the effect of securing the car to prevent its tipping.

These were powered by batteries for the first two years of the line's operation. In the final year, a trolley wire was installed. Power to charge the batteries, and later to energize the trolley wire, came from a generator connected to a stationary steam engine powered by coal or wood located at the Crystal Beach end of the line.

Departures from Crystal Beach were announced by whistle blasts from the steam engine in the power plant.

Filling the seats of the monorail was a challenge. Builders of the line counted heavily on commuters between Crystal Beach, Ontario and Buffalo, NY who could take the Grand Trunk trains to or from Ridgeway Ontario.

The system nominally operated service at 15-minute intervals, however its reliability was poor. The posts would sway in soft ground, especially after heavy rains, and winter frosts shifted the posts, necessitating substantial maintenance each spring. In spite of the maintenance challenges facing the railway, no major accidents are known.

There were two documented accidents. The first was a fatality to a cow. It was struck by one of the 55 battery cases which hung beneath the car. Another time, part of a woman passenger's basket of laundry was sheared off at the underpass.

A lack of passengers made the line uneconomical. Following the 1898 season, the railway was closed and dismantled.

What to see? The Clark family of Ridgeway Ontario had a very distinct history with the long defunct monorail. Judge Joseph Clark had worked on maintenance and as motorman for the line. Another Clark worked at the line's power plant.

Joseph Clark's son Frank built a scale model of the roadway. It was donated to the Fort Erie Museum. It is in storage. It would be prudent to call the museum first before you visit to see how long it would take for them to get it out of storage for you to view.

Canada Southern Railway Car. credit Fort Erie Public Library

NIAGARA FALLS MONORAIL PROPOSAL 1963-2009

1963 - 1971 - Goodell Monorail (Niagara Monorail)

The dream of having a monorail system in Niagara Falls began in 1963. Monorail Incorporated from Houston Texas proposed having a monorail run between Niagara Falls, Ontario and Niagara Falls NY.

Monorail Inc. was run by Muriel Goodell, whose monorail systems he set up in the 1950s in Houston and in Dallas Texas. His proposal was contingent on permission from the Niagara Falls Bridge Commission to use the Rainbow Bridge to cross the river.

Goodell Monorail in Dallas Proposed for Niagara Falls. credit author's collection

Engineers from the company had been in the area for several months surveying the entire route and researching all the area for suitable facilities. The type of monorail suggested for Niagara Falls would be the Monorail Inc. model they operated at the Dallas Fair.

By January 1964 the proposal had hit a snag as the Niagara Falls Bridge Commission had requested more information from Monorail Inc. before they would consider approving the idea.

The cost was estimated at $12 million and fare cost would be $1.50.

According to a New York Times article, on February 18th, 1965 Niagara Falls NY council voted 4-1 to grant a 20yr franchise to Goodell Monorail Inc. which will build the monorail 30ft above Falls Street. The monorail had to operate at least six months a year. The city also had the option to order the removal of the monorail on 90 days' notice. The $2 million, 10 block monorail would run from downtown Niagara Falls, NY to Prospect Point. Adults would be 75 cents, and kids under 13 would pay 50 cents.

The idea to have an international monorail was abandoned.

On the Canadian side a monorail proposal made headlines on July 23rd, 1968 The proposal by Monorail of Niagara Ltd. (a Goodell Monorail subsidiary) would run between Seagram Tower and the New York Central right-of-way, extending from behind the tower southerly toward Livingston St.

The initial plans called for the monorail to extend for 400 ft. A Salt Lake City investor would put up $100,000 for a one car demonstration test project. The demonstration line was built and a small demonstration monorail car was used.

The company's single car monorail system that was then at Heritage Tower would bear little resemblance to the coaches planned for the new system.

The only known photo of the demonstration monorail line right side of photo, February 1999. credit Mike McIlwaine

A July 18, 1969 Niagara Evening Review article stated that Niagara Monorail Ltd was announcing plans to extend the proposed monorail to the foot of Queen St. Niagara Monorail noting that the cost would be $12 million. Niagara Monorail had paid Penn Central $900,000 for air rights which will allow them to construct two single rails above the tracks.

A newspaper article on December 2, 1969 revealed that the monorail plan by Niagara Monorail Ltd to link Queen St and Chippawa St had gained the support of city council. Council wanted the monorail system to be conceived as a solution to mass rapid transit in the city.

The main hold up was the city's inability to authorize operation or construction of a monorail over city streets. There was no existing provincial legislation that allowed for the building of a monorail. Additionally, a city report outlined several more hurdles that would have to be resolved before the monorail proposal could move forward.

Niagara Falls Mayor Franklin J. Mayor couldn't hide his unhappiness with the situation. He stated, "I know we have bylaws and regulations but I'm disappointed to find nothing or little in these reports to encourage erecting a monorail in this city."

The developer's lawyer Judy LaMarsh was already in talks with the provincial government about a bill to allow the monorail to be built.

Daily News (New York, New York) on 7 Jan 1970, stated that the federal government's Canadian Transport Commission gave Niagara Monorail Ltd. authority to build a 4.35 monorail system along the Niagara River at Niagara Falls.

The government of Ontario passed Bill C-157 The City of Niagara Falls Act, 1970 (No. 2) into law May 1970. It allowed the city the right to build a monorail.

With the federal and provincial government approval in place there was optimism the monorail would be built.

Clearing of land starts for monorail June 5th1970
- credit Niagara Falls Review

On June 5th, 1970 a Niagara Falls Review article stated that construction had started clearing the way for the monorail route. Fred Ingham Construction Co. Ltd of Niagara Falls would be clearing the route along the side of the Penn Central railroad tracks.

The line would carry tourists and commuters 23 feet above the railway tracks six miles from Chippawa to downtown Niagara Falls with stops along the way. The entire tour would take 22 minutes with one-minute stops along the way.

There would be 12 trains, each seating 200 passengers running during the tourist season. It could transport 14,000 persons hourly.

1970 Monorail groundbreaking plaque.
Photo credit - Niagara Falls Railway Museum

Niagara Falls Monorail put out a commemorative 1970 Monorail groundbreaking plaque that was given to some people.

Niagara Monorail was a subsidiary of Goodell Monorail of Salt Lake City, Utah. As mentioned, Muriel Goodell had paid Penn Central $900,000 for rights to run the monorail above their track. In 1970 Penn Central went into bankruptcy thus creating even more financial stress for the cash strapped Niagara Monorail.

A July 9th, 1970 Niagara Falls Review article stated that Penn Central had wanted to do some work on their track and it delayed construction of the monorail during June and July 1970.

Modern Steel and Iron Works were working on the prototype coach for the system. They halted work in July 1970 with 50% of the coach completed.

These delays also caused further hardship for Niagara Monorail.

On January 9th 1971, the Niagara Falls Review headline read, Monorail may be abandoned. Muriel Goodell, the president of the firm to build the monorail in Niagara Falls, stated that financial problems could cause the death of the project. Goodell Monorail, was suspended from being sold and traded on the Salt Lake City, Utah, Stock Exchange.

A financial statement from December 21, 1969 stated that Goodell Monorail had assets of $113,000 and liabilities of $550,000. Nonetheless, Muriel remained upbeat about the future of the monorail being built. Unfortunately, the reality was this chapter in Niagara Falls monorail dreams had come to an end.

1996 - 2009 - People Mover System - Monorails, Light Rail or Buses

A March 1996 the Niagara Falls Review article outlined a new $140 Million monorail idea. It was the result of a two-year long feasibility study by Totten Sims Hubicki Associates that looked into a monorail type transportation system.

The system could shuttle about 3,600 people to about 10 destinations every hour. The trains would make a one-way 10km loop following the Niagara Parkway and a CN Rail corridor atop Murray Hill.

The 1996 propane powered People Mover seasonal buses, used since 1986, carried about 1.2 million people a year. The proposed monorail could move almost eight times as many people.

Financing and operation could be a public/private partnering. Trains could be running by the year 2000. The next step was the preliminary design and environmental assessment process.

A June 1997 article in the Niagara Falls Tribune stated the idea of a People Move System was moving along and the proposed terms of reference for studying the environmental impact were revealed.

The existing bus travelled at an average speed of 4km/h and needed to be upgraded. This article did mention monorail as an option but unlike an earlier article started using the term People Mover, which could mean more buses.

By June 23, 2005, The Niagara Falls Review article had moved away from the monorail idea to a ground level light-rail time system.

Ed Lustig, the People Mover System's Director, stated the ridership would be the same if the system was elevated and or ran along the ground. He stated that the People Mover System favours some kind of light-rail running at grade.

It would run at grade along the former Canadian Pacific Railway corridor from Casino Niagara to Marineland, then along the Niagara Parkway to the Maid of the Mist.

The Parks Commission was strongly in favour of an at-grade system.

On April 24th, 2009 Niagara This Week - St. Catharines news title said it all, "City Bids Farewell to Monorail - People Mover to come in the form of buses." A reprint of the article below explained the situation.

"The plan for a monorail in the city in the form of the Niagara Falls People Mover has been stalled on the tracks.

Council gave former CAO John MacDonald the go-ahead Monday to move forward with a people mover system by bus operated by Niagara Transit, as opposed to a monorail idealized for decades.

The rush is on as the deadline to submit an application for $25 million in grants from the federal government is due in June. The city's people mover — which is different from the current system of the same name operated by Niagara Parks Commission — has been halted for years. Stakeholders like Niagara Parks and Falls Management Company (the company which runs the casinos), the Ontario Lottery and Gaming and Niagara Parks have been working on the project for a decade and millions of dollars have been spent on studies.

The city's people mover would replace and expand the Parks Commission's People Mover system, as well as Niagara Transit and hotel shuttles. It would be city-owned and operate on city streets and in Queen Victoria Park, provide service to most major attractions including the new convention centre set to open in 2011, and link

Lundy's Lane and NPC attractions north of Victoria Avenue. There would be year-round and seasonal routes.

MacDonald estimates upwards of 30,000 riders on peak days and identified casino-to-casino ridership as "important."

Phase one of the project will be on tires and phase two, a fixed rail system, would include a new RFP process in the private sector after council's decision Monday dropped three companies previously eyed by council to operate the preferred rail system.

Whether the system ends up being viable or not, the city can't lose because it will receive $25 million in the form of new buses, said MacDonald.

For Coun. Wayne Thomson, who has long been an advocate for a rail people mover, any system is a good one.

'After 20 years of disappointments at this point, everyone has attempted to pull this thing together," he said. "The rubber tire system is a totally inappropriate situation but it may be the only thing we have.' "

Are there any other monorail related stories in Niagara Falls past?

In February 1998, Falls Management Co consortium was selected to finance, design, build and operate what would become the Niagara Fallsview Casino, which opened in 2004. In the original proposal a monorail to connect different areas was proposed. It was not built.

Marineland Monorail Track - There were monorail beams on the ground next to the storage area at Marineland. Perhaps it was bought by Marineland owners to build a monorail. It never was built and the beams are now gone.

Is there a future for monorail in Niagara Falls?

The City of Niagara Falls Sustainable Master Transportation Plan by AECOM was released in 2011. It promoted biking, buses and road widenings.

It is doubtful that after 60 years of trying, a monorail will ever be built in Niagara Falls.

TORONTO 1954 - 2014
MONORAIL PROPOSALS

The City of Toronto has a long history of saying no to monorails. The first attempts to get a monorail built in Toronto started in 1954.

Like in other Canadian cities at that time, they were interested in the Goodell Monorail. Muriel Goodell and his Houston based Monorail Inc. company had built a prototype operating monorail in Texas and cities were looking at it as a way to solve transportation problems.

Toronto 1954 - 1959

The Toronto Planning Commission investigated monorails in 1954 and rejected them as an alternative to subways. Nevertheless, the 1950s saw repeated requests for considering the monorail for Toronto and the surrounding area. Here are some highlights.

An editorial on September 21,1955 in the Toronto Daily Star, suggested looking at a monorail line running parallel to Bloor Street as an alternative to subways. The article mentioned that Houston and Montreal were studying monorails.

On March 27, 1956 an article in the Toronto Daily Star stated that the Toronto Transit Commission Chair Allan Lamport went on a fact-finding mission in Texas to ride the Goodell monorail demonstration line. He rejected the monorail as an alternative to the subway based on the capacity limitations of the monorail.

Goodell Monorail Test Track Texas. credit author's collection

In 1958 the idea of having a monorail system in Toronto again was raised. The Reeve of North York Vernon Singer and the Reeve of York Township Christopher Tonks pushed to investigate the idea again.

Reeve Singer wanted a deputation to be sent to Texas to look at the Goodell monorail before any decision on the subway was made.

Reeve Tonks and Reeve Singer did visit the monorail system in Texas and were supportive of it being built for Toronto. Allan Lamport was totally against the monorail.

On May 3rd, 1958 the Toronto Daily Star ran an article stating the president of Monorail Ltd. (Muriel Goodell) was cancelling his meeting with Metro Council. He wanted more information on Toronto's rapid transit program to study it.

Originally Singer and Tonks wanted to meet with Goodell in Singer's office. This was viewed negatively by the Metro Council. The cancellation by Muriel Goodell to attend the meeting further supported those against the project like Metro Toronto Chairman Frederick Gardiner and Allan Lamport.

A week later Metro Council did approve by a slight 9-8 margin for a study on the feasibility of a monorail system in Toronto.

Avro Aircraft Division to Study Monorail.

In a move that surely showed anti-monorail bias, Frederick Gardiner commissioned A.V. Roe's Avro Aircraft division to study the use of monorails in Toronto! Even more suspicious was according to the Toronto Star on June 21st, 1958, when Gardiner announced the findings of the report, he kept secret who did the report.

On the basis of that secret report, Metro Council voted against the study of monorail. A week later it was revealed that A.V. Roe did the report. Though it stated there was no place for monorails within the city limit it might be suitable for the suburbs. The route considered was the CNR right of way from Union Station to Malton Airport.

By May 26th, 1959, the Toronto Star reported that even with the AVRO President supporting the idea of the Union Station to Malton Airport, the four suburban Reeves had a tough time getting support for the idea.

The idea of the monorail to the airport did not progress.

One other monorail idea in the 1950s is worth noting. On May 21st, 1959 the Toronto Star reported that Etobicoke township was given a presentation on monorail. Reeve H.Q Watts stated it was a feasible practice to run a monorail along the Queen Elizabeth Way from Toronto to Hamilton. This idea went nowhere also.

Toronto 1960-1969

The 1960s monorail dreams for Toronto focused mainly on two ideas for monorails at the CN Exhibition.

The decade started with a Quebec engineering firm Milot and Sond Inc. of Jonquiere, Que suggesting the monorail system they had designed. They stated it could be an alternative solution to transit extensions such as the Spadina expressway. That idea went nowhere.

In 1964, The Toronto Star reported on June 3rd on the idea to have a monorail run from the CN Exhibition grounds to Toronto Island.

In January 1965, The Toronto Transit Commission scrapped the idea of a full scale monorail as too expensive. The CNE centennial committee still was researching the idea to have a smaller monorail (skyride) built that would just run on the Exhibition grounds in time for the 1967 celebration.

By 1996, the cost for the Sky-Ride monorail ballooned from $1.4 to 2.6 million resulting in the cancellation of the idea.

The decade ended with a plan from businessman Douglas Wells in 1969 to revolutionize the CNE. It included a two-level $1 million grandstand stage. He planned to build picturesque islands in the lake with a steel-and-concrete parking garage under them. Access to the islands would be by monorail and covered walkways. Not surprisingly, this scheme failed also.

Toronto 1970-1979

There were not many monorail dreams in the 70s Toronto. Here is a summary of a few of the ideas then.

An Ontario government task force started off the decade with a 1971 report by experts that outlined what future transit might look like in Toronto and the other places in Ontario.

The task force report did mention monorails and was very supportive of the France Aerotrain, which rides over a beam on a cushion of air.

Then on November 30th 1971 the Toronto Star reported an update on the 1968 Canadian National and Canadian Pacific Railway $1 Billion Waterfront Renewal Project called Metro Centre. One of the aspects of the plan included a German designed monorail.

In 1971 the Ontario Government got involved in modern automated guideway transportation systems and sent out a tender for interested companies. Although the term "monorail" was used in the newspapers at that time to describe some of the vehicles being considered, none were monorails.

Krauss-Maffei Transurban, not a monorail. credit author's collection.jpg

The winning entry was a German made magnetic-levitation vehicle by Krauss-Maffei Transurban. Though maglev vehicles could be designed like a monorail, this vehicle was not. In 1974, funding for the Krauss-Maffei maglev was pulled by the German government and their project with the Ontario government was cancelled.

In 1972 the idea was floated around that a monorail would be part of the proposed $2 billion Pickering Airport project.

In 1973 another misuse of the word monorail was used to describe the new automated guideway people mover being installed at the Toronto Zoo by Bendix Systems Canada. The term monorail was commonly used to describe the zoo vehicles for decades. They definitely were not monorails.

The last reference to monorails in the 1970s in Toronto was the report from Richard Soberman, director of the Metropolitan Toronto

Transit Review, whose 70-page master blueprint for transportation in Metro mentioned monorails.

Toronto 1980-1999

The 1980s and 90s were a quiet period for people/companies promoting monorails.

You will remember Lyman Richardson, whose TranSyt monorail company tried to get a test track built in Calgary during the 1980s. The North York company was promoting the speeding hanging bullet monorail capable of 240m/h, in North York.

TranSyt Monorail Model - credit source unknown

Richardson had the small-scale model vehicle on display at North York City Hall in May, 1982. In August of that year, he met with Mayor Lastman, Metro Chair Paul Godfrey and the TranSyt bank backers from Arabia at the North York Civic Centre. A TranSyt monorail did not happen in North York.

Over in Mississauga, the Toronto Star reported on Tuesday, March 2nd 1982 that the future of Mississauga would have hotels, office towers lined by a monorail. This idea was never mentioned again.

In 1996 there were 3 proposals vying to be picked for Ontario Place and Exhibition Place renewal. Toronto owned the 105-hectare Exhibition Place site and the province owned the 44-hectare Ontario place. They would retain ownership. While the successful bidder leased the land and buildings. The proposal from Agra Foods Inc., a division of Apotex Inc. called for a year-round theme park, with hotels, restaurants, shopping and an internal monorail system. The monorail was not built.

Toronto 2011-2014

On August 11th, 2011 the National Post ran a story stating that the City of Toronto wanted to take the reins of redevelopment plans for the industrial port lands - prime waterfront real estate that Councillor Doug Ford was keen to transform into Toronto's next great destination.

Waterfront Toronto, a tripartite agency funded by the municipal, provincial and federal government had been overseeing port land redevelopment for ten years.

Doug Ford's early vision included an elevated monorail among other things like a giant Ferris wheel. Councillor Michael Thompson was open to the monorail idea. The idea was supported by his brother Rob Ford.

By August 31st, 2011, a map outlining Doug Ford's plan for the waterfront - which included a mall, hotel, Ferris wheel and a monorail route - was on page 1 of the Toronto Star.

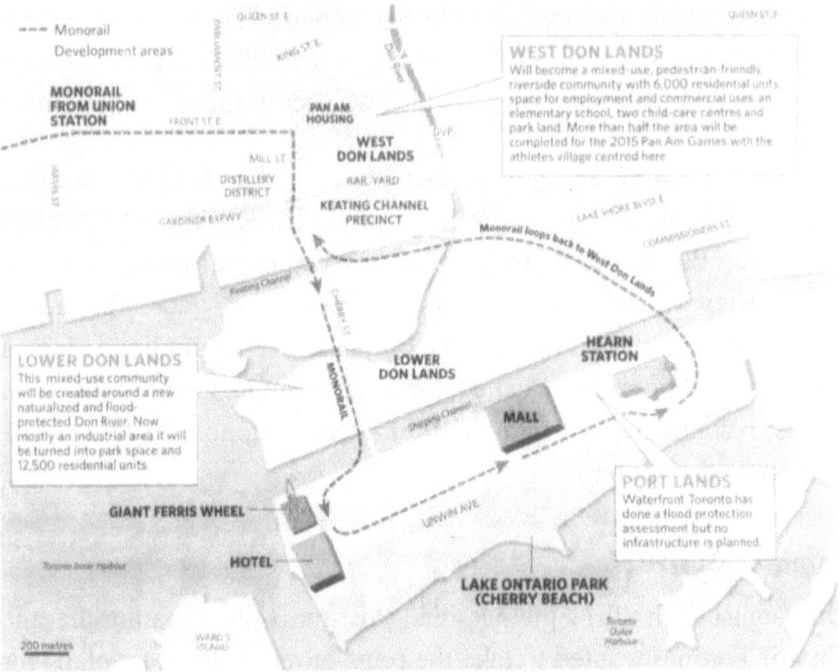

2011 Doug Ford Plan. credit Brian Hughes Toronto Star

On September 1st, a Toronto Star editorial was negative toward the Ford plan and monorail. On the 5th of that month the Star did an article on monorails featuring the Monorail Society founder Kim Pedersen.

The ambitious plan by Doug and Rob Ford to gain control of the port lands for their development plan faced opposition from many professionals, the media, and some councillors.

By the middle of September, it seemed clear that the Ford brothers' plan was in serious jeopardy and would not be approved at the September 21st Toronto council meeting.

Sure enough, at the council meeting the proposal to take over control of the port lands from Waterfront Toronto was defeated in a unanimous vote. A stunning turn-around from two weeks earlier when the executive committee had unanimously backed it.

Thus, ended Toronto's latest attempt at having a monorail.

MONTREAL MONORAIL PROPOSALS 1947-2020

The research for this book on Forgotten Monorails of Canada left no doubt that Montreal and the surrounding area were the leaders in studying monorails in the last 73 years. They do have two monorails running in the province, one left over from Expo 1967 at La Rhonde Amusement Park and the other at the Granby Zoo. These will be examined in separate chapters. This chapter focuses on the ideas/studies/schemes for inner and inter-city monorails relating to Montreal that never became a reality.

As far back as 1947 Edward Hanson, vice president of Gibbs Hill Inc. was promoting the idea of a monorail system for Montreal at the weekly meeting of the Engineering Institute of Canada. In a November 7th, 1947 article in the Montreal Gazette, Hanson stated, "An overhead monorail system whose cost would be about 20% of that of a subway was seen as the immediate solution to Montreal's traffic ills. The suspended monorail in operation in Germany since 901 with modifications, is a possible answer to traffic problems." Further discussions were held in Montreal in 1948 about monorails but nothing moved forward.

1953 - Monorail Instead of Subway or Expressway Proposal

Montreal to Study Monorail System for City; Cost $7,500,000 Less Per Mile than Subway

Montreal Monorail Conceptual August 23, 1953. credit Montreal Gazette

On August 22 1953, the headline of the Gazette paper was “Montreal to Study Monorail System for City; Cost $7,500,000 Less Per Mile Than Subway.” Waiting for the release of the Montreal Transport Commission's long-awaited subway report, city representatives stated they were going to study the overhead monorail. To facilitate consideration of a monorail's desirability and potentialities, the city planned to send representatives to Los Angeles for information concerning a monorail.

On August 29th, 1953, approval was given to Montreal city officials to visit monorail plans in various USA cities. Looking at this from today's perspective this was not a great idea. There were no monorail systems in operation to see and the USA were in the midst of transforming from street car based transit systems to massive highway building. The politics and big business interests that eliminated local inner city rail transit would also keep monorail systems out of urban centres.

An article on September 19th 1953 in the Gazette detailed the visit of a Gazette reporter to Wuppertal Germany to visit the operating monorail there. The reporter stated that the monorail runs at a deficit and did not replace tramways in the city that run parallel to it. The reporter warned Montreal officials to think twice before choosing monorails. Again looking at it from today's perspective it was a good idea to ride an existing system in operation. However, the reporter did not consider the fact he was riding on a system that was half a century old.

The Municipal Congress was held on September 22, 1953 and some of the mayors of USA cities questioned why Montreal officials were visiting the USA cities stating monorails are a second-rate solution to transit problems and do not exist in the USA.

In a March 23rd, 1954 Gazette article, Montreal Transportation Commissioner C.A.Sylvestre, stated, "Only a subway can solve Montreal's traffic problem and it will be built when enough people want it built." He based his comments on the subway report which was submitted to city council in October 1953. Mr. Sylvestre attacked elevated railways, monorail systems and expressways. " The districts surrounding the pillars of an elevated railway would become a slum. More and more American cities are replacing their elevated systems with subways."

Editor's note: Not sure where Mr Sylvestre was talking about as there were no monorail installations in the USA in 1954.

An editorial on the monorail appeared in the Montreal Gazette, Friday Sept. 16th, 1955 titled, "This Seems a Little Far-Fetched". The editorial was correct in stating that it was unproven technology. Only the Wuppertal Monorail in Germany was in operation and the USA pioneering monorail company Monorail Inc had just started its testing of their Trailblazer monorail on their test track at Arrowhead Park, Houston Texas. The lengthy editorial ends with the statement that, "It hardly seems wise to rush into the unknown in hope of clasping a solution."

Sept. 15th, 1955, the Gazette reported that Montreal Planning Director C.E. Campeau proposed that "serious consideration" be given to the monorail system of mass transportation in Montreal. He gave detailed figures of how much cheaper monorails were to build compared to highway building and subways. He also talked about how much less space is needed for a monorail than an expressway. He asked the city council to be authorized to visit Houston Texas to ride the Monorail Inc. monorail on the test track in Houston.

At that time, Muriel Goodell the president of Monorail Inc had several Canadian cities interested in the monorail system. A 1956 film of the Houston monorail tests mentions several Canadian and USA city officials riding the monorails and shows footage of the full monorail cars carrying the guests.

By 1961, this monorail dream for Montreal was over. An article on January 24th in the Gazette stated that the subway was the top priority for the city. The report presented by the Montreal Metropolitan Corporation recommended to the city technical subcommittee that the entire 14.7-mile initial phase be underground. After investigation the subcommittee reached the conclusion that a monorail would not be practical for Montreal. By 1962 work began on the subway.

1962 - Montreal to Dorval Airport Monorail or Subway to Airport Talks

In December 1962 it was reported in local papers that the federal government and the City of Montreal officials had discussed the possibility of extending Montreal's subway to the Dorval airport. Transport Minister Leon Blacer and Mayor Jean Drapeau also discussed the alternative of building a monorail between Montreal and the airport 15 miles away. Nothing became of these talks.

Monorail in Ad December 30th, 1963. credit Montreal Gazette

1966 - Montreal to Utica NY Rapid Transit Monorail Proposal

On May 11th, 1966 the Gazette and other papers reported that a group of New York state business men were promoting a scheme to build a $600,000,000 monorail from Montreal to Utica. Gary Lytle, a Utica warehouse owner headed the group. The plans included using the New York Central railway right-of-way as a base on which to build the line. He estimated the travel time would be 3hrs between

Montreal and Utica NY. Lytle promised to meet with Montreal officials after receiving a feasibility study.

June 2nd, 1966 the Gazette reported that top level business executives from northern New York met in Malone NY to discuss the projected Montreal-Utica monorail link. The group pushing the idea had suggested that the Canadian, USA or even New York Central Railway who owned the right-of-way would pay for the cost of building the line.

The existing passenger rail service on the line was losing money. New York Central was open to leasing the space needed for the monorail pylons.

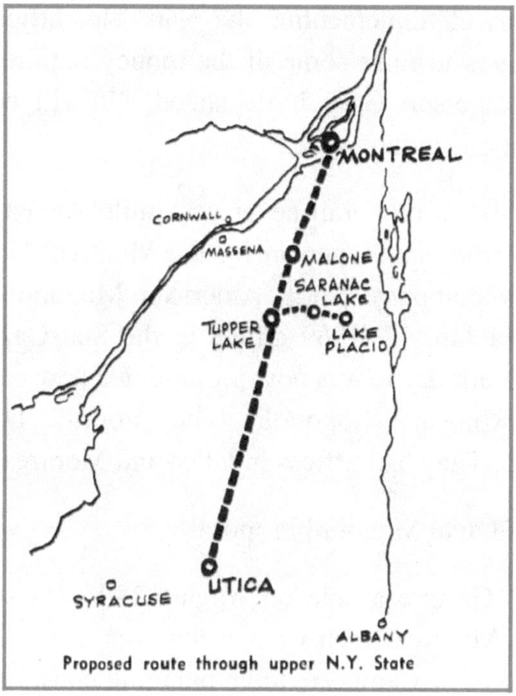

Montreal to Utica Monorail map May 11th, 1966. credit Montreal Gazettte

On December 8th, 1966 - The Utica N.Y. supporters of the monorail system met in Montreal at the Windsor Hotel for local business in Montreal interested in the idea. The Utica group was now named the Mohawk Valley-Adirondack Montreal Monorail Inc. The cost of the project was now going to be $1.5 billion to complete. My Lytle spokesperson for the group had said that the Transportation Systems Division of General Electric Company had promised to provide technical advice and information to the group. General Electric had recently purchased a franchise from Safege-Transport in Paris for use of that monorail technology.

The Gazette reported on June 13th, 1967, that backers of the Montreal - Utica monorail link had placed the project in the hands of top engineers and economic consultants who were studying the most practical means of implementing the plan. This article stated they would sell shares to raise some of the money required to fund the project. Spokesperson Gary Lytle stated, "It will be one of the greatest attractions in the world."

The Montreal-Utica monorail never was built. Amazingly, in 1969 Gary Lytle, former spokesperson for the Montreal-Utica monorail, headed a new company North American Monorail Corporation. According to a Jan. 17, 1969 article in the Star-Gazette paper of Elmira, New York, Lytle was now promoting a new concept vehicle called the Rotocar Monorail, 'The Fourth Dimension of Transportation. They had offices in Utica and Montreal.

1968 - City of Laval Monorail Proposal

The Montreal Gazette article on August 28th, 1968 stated that a monorail and Montreal metro extension was needed. A 160-page report by Charles E. Campeau, town planning consultant, called for a 26.9-mile-long monorail that would be linked to a new extension of the Montreal Berri-Henri Metro line to St. Martin Blvd. in Laval's industrial park.

On January 25, 1989, twenty-one years later, The Mayor of Laval Claude Lefebvre was very supportive of monorails. He unveiled plans for a monorail and had a small-scale model of one.

On August 28th 1989, the Gazette reported that Premier Robert Bourassa promised a transit link between Laval and Montreal. The options of expanding the metro. The idea of the monorail was still being promoted by the Parti du ralliement officiel des Lavallois.

No monorail system was ever built. The metro was extended to Laval, the Cartier station opened in 2007.

1968 Expo Express cars modified to run as monorails idea

On September 14, 1968, the Gazette reported an interesting concept. Some well-known anonymous Canadian businessmen were going to buy the Expo Express transit system cars and convert them so they could run on a monorail. The Expo Express used standard railway technology (two rails) and a third electrified rail. It was a fully automated technology. The Expo Express cars were not sold to this company, so it was never proven if these cars could have been converted to a monorail style.

1968 - City of Anjou Monorail Proposal

Two days after the City of Laval monorail the City of Anjou stated they also wanted a monorail. Mayor Ernest Crepault stated in the Gazette, "Creation of a rapid transit system is necessary to realize the municipality's potential." The city hired the same consultant Charles E. Campeau who did the study for Laval. No monorail system was ever built.

1969 - Montreal to Ste-Scholastique (Mirabel) Airport Monorail Proposal

It was reported on June 9th in the Montreal Gazette that the federal government was considering using the "Aerotrain" to connect Montreal to the airport. The Aerotrain was a jet engine powered monorail that ran on a cushion of air capable of speeds of 368 km/h. During 1969 it was still in the testing stage and footage of the test runs were impressive. Jean Luc Pepin, the Industry, Trade, and Commerce Minister, speaking informally at a Paris airs show, stated that he was considering that system among others. The France company La société Bertin (the inventor of the Aerotrain was Jean Bertin) had set up a consulting firm in Montreal with a sale of the Aerotrain as a goal.

Aerotrain France. credit author's collection

On October 30th, 1970 the Gazette reported that the company EMS Electronik Monorail Sales was proposing that a double track straddle beam monorail, like the system used in Seattle, was the best for the Montreal to Mirable link. Electronik Monorail Sales was the Canadian agent for monorail company Rapid Transport Development. A decision by the government of the preferred technology (30 firms submitted proposals) was expected to happen in 1971.

On March 8th, 1973 the headline for the la presse newspaper was asking which technology was going to be used to connect Mirabel to Montreal, Monorail, Aerotrain or Metro. No transit link was ever made to connect Mirabel Airport to Montreal. It stopped having commercial passenger flights in 2004.

1972 - Ste-Scholastique (Mirabel) Airport to Dorval Airport Monorail Proposal

On August 4th, 1972 a study released by the Dorval Chamber of Commerce stated that shutting down Dorval's airport would sever a vital economic lifeline for the community. The study suggested Dorval airport stay open and connect to Ste-Scholastique Jetport by means of a rapid transit monorail.

1974 - Montreal Transportation Study Announced

On April 11th, 1974, the Gazette reported that Quebec Transport Minister Paul Berthiaume indicated that no new bridges will be built until a fully-integrated public transit was developed. A $1 million study would begin in the fall outlining transportation south shore needs. The study would last two years. He doubted the use of the "exotic" monorail or other types still in the experimental stage.

1987 - Transyst Canada Monorail Proposals

Transyt Canada based out of Toronto had tried unsuccessfully to build a test track in Calgary for their monorail system during the period of 1982-1984. Their president Lyman Richardson first pitched their system in Quebec to connect Mirabal and Dorval airports. When that option was too expensive he proposed that the monorail could be attached to the Jacques Cartier bridge or follow the ice dam west of the Champlain bridge. Mr. Richardson said, "At a capacity of 24,000 people per hour, the monorail could easily handle all the traffic of the Jacques Cartier, Victoria, and Champlain and Mercier bridges combined."

1992 - CN/AMF Rail MonTrain to South Shore and Laval Proposal

December 1st, 1992 - CN Rail revealed a mass transit proposal that would see Montreal-area commuters travelling more quickly and cleanly via monorail and new commuter trains. AMF is an independent division of CN specializing in the manufacture and reconditioning of transportation equipment. The proposal, called MonTrain, would include construction of an automated electric monorail linking Brossard and Nuns Island to Central Station via the South Shore ice-bridge. The plan also called for an above ground monorail line along Hwy 15 through Laval. If implemented, the MonTrain plan would create 2,000 manufacturing jobs.

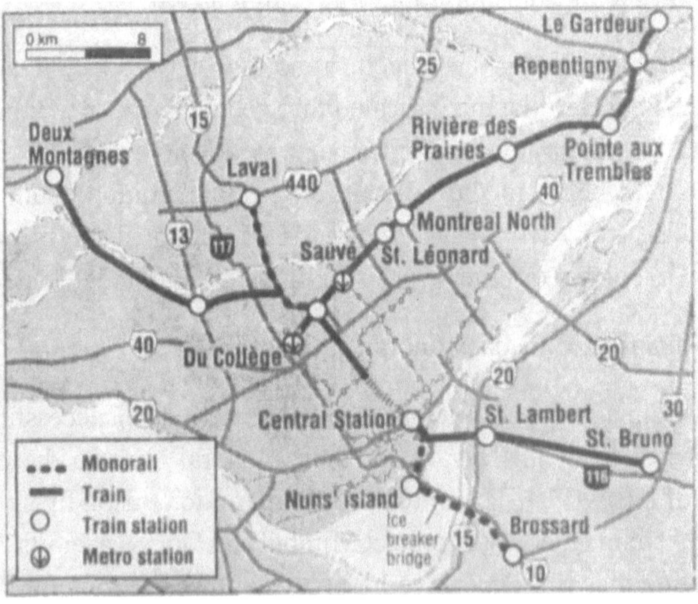

CN Proposal Map Dec 1, 1992. credit Montreal Gazette

MonTrain also included plans or a South Shore train using existing CN Rail lines. The plan would cost $460 million approximately. This plan was put forward as CN-CP could not reach a joint agreement on the previous CP Rail plan in February 1992.

The CP Rail plan called for development of 365 km of existing CP and CN track. It was patterned on the GO Transit commuter system used in Toronto. CP would use refurbished GO trainsets if Quebec purchased them. CP proposed their commuter plan will be run by a special wholly owned subsidiary of CP Rail. CP has estimated that passenger rail revenues would cover 38% of the costs.

By June 1993 the Gazette had reported that CN and CP had come to an agreement on their commuter rail plan. It was based mainly on CP's original plan. The monorail was not in the revised plan but was still important to AMF as it meant local jobs. The province had until the end of the year to buy 80 GO transit rail cars from Ontario at $4.9 million. Both CP and CN said the deal was too good to pass up.

It is interesting to note at this time that there were also two proposals to use an ice-bridge near the Champlain Bridge and turn it into a $71 million bus route.

The 1994 Daimler-Benz AMF "Sprint" Monorail Proposal

Monorails were again suggested as a solution for Montreal's South Shore transit link. Daimler-Benz with AMF unveiled their "Sprint " driverless rapid-transit trains proposal. A Montreal Gazette article dated September 8, 1994 stated that it would link Central Station in Montreal and a terminal in Brossard. Enroute was a stop in a new real estate development on Nuns Island. It then followed an elevated guideway along the ice-bridge near the Champlain Bridge. This proposal would be competing directly with the bus proposals mentioned above.

The trains would run on rubber tires on steel tracks along the sides of the guideway, with lateral control provided by horizontal wheels locked into a central I-beam track. The mayors of Brossard and Verdun have endorsed the proposal.

1995 Consultants Canarail Canada Inc Study

On October 5th, 1995, the Gazette reported that a consulting firm, Canarail Canada Inc. completed a study that was released by Quebec Transport Minister Jacques Leonard. It recommended moving the buses from the reserved bus lanes of the Champlain Bridge to an adjacent ice-bridge. The cost to shift the buses to the ice-bridge was estimated at $68.4 million. The study did not support the joint CN/CP commuter rail plan or the Daimler-Benz Sprint monorail.

2000 Roche Deluc Studies

It was reported in the Gazette on January 13th, 2000 that the Montreal engineering firm Roche DeLuc in one study called for $600 million to be spent on two new bridges and an extension of Highway 30 linking the South Shore at Beauharnois with Vaudreuil/Dorion. In a second study it recommended an "elegant" $300 million monorail crossing the St. Lawrence on the existing Champlain Bridge ice control structure. It would have a stop at the casino.

The monorail cars would be built locally by Bombardier and the monorail link would initially carry up to 14,000 passenger cars an hour in each direction with an ultimate capacity of 24,000 passengers per hour.

Politics came to the forefront when the Quebec Transport Minister Guy Chevette complained in a Gazette article on January 14th, 2000 stating the two studies into Montreal's traffic woes, commissioned by the Jacques Cartier and Champlain Federal Bridge Corp. are more federal intrusion into provincial jurisdiction. However, Mr. Chevette's comments were not supported by all provincial authorities. The Agence Metropolitaine de Transport, a provincial agency which oversees public transit in the metropolitan region had offered the studies' results. AMT director Florence Junca-Adenot had no problem with the agency's participation.

2002 Loto-Quebec New Montreal Casino, Hotel, Restaurants and Monorail Proposal

In November 2002, Loto-Quebec announced a $470 million project to convert the former Quebec Pavilion, across from the building currently housing the Montreal Casino. It would include a new casino, restaurants, hotel, and a monorail linking the casino to the island of Montreal. However by October 13, 2003, according to an article in the Gazette, The Quebec Finance Minister Yves Sequin had slammed the door on the opening of new casinos in the province.

An article in the Gazette June 23rd 2005 showed a map of the Loto-Quebec proposal and mentioned that Cirque du Soleil was enlisted as a creative partner in the plan. By February 2006 Loto-Quebec had abandoned their proposal and the proposed monorail died with it.

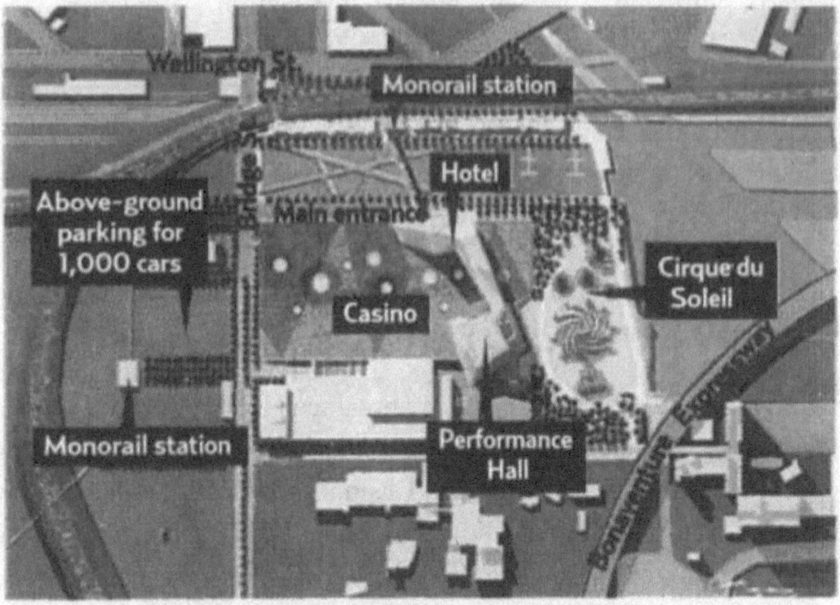

Planners envision a monorail connection, boat basin at casino.

SOURCE: LOTO-QUÉBEC/CIRQUE DU SOLEIL

In 2007 Agence métropolitaine de transport proposed a light rail route using the ice-bridge structure next to the Champlain Bridge. No action happened on that proposal.

In 2007 the process began to replace the Champlain Bridge. The Gazette reported on April 11, 2013, that the federal government pledged to support the building of the light rail on the new Champlain Bridge.

Finally, in 2015 construction began and the bridge was opened in 2019.

In 2016, Réseau express métropolitain (REM) light rail 62 km system was approved to be built. It will connect the South Shore communities, as well as provide a long awaited link to Pierre

Trudeau Airport, among other routes. The new Champlain Bridge will have two lanes for the REM.

2008 Montreal-Quebec City High Speed Monorail Proposal

Trens Québec monorail. credit TrensQuébec website

Quebec physicist Pierre Couture. Dr. Couture, a Princeton University graduate spent most of his career working at Hydro-Québec, where he headed a team of researchers at TM4 Inc., a subsidiary of the Crown corporation. Between 1982 and 1995, working with a small budget, the team developed an electric-powered motor-wheel. In what is still a controversial move Hydro-Quebec dismantled half of Dr. Coulture's team.

Mr. Couture had developed a high-speed rail suspended monorail based on this technology. A group formed with Jean-Pierre Marchand, director TrensQuébec, and Pierre Langlois, physicist and specialist in sustainable mobility to promote this technology.

The idea was to run a high speed suspended monorail from Montreal to Quebec-City and other locations. It could run at 250 km/h. According to the TrensQuébec website, it would cost less than $12 million per km, which is about three times less expensive than a TGV line.

The proposal got a boost when, at the Liberal Party convention in November 1917, Premier Philippe Couillard launched the idea of developing "a modern, sustainable, futuristic link" between the two largest cities in the province, thereby opening the door to the monorail.

The proposal faced opposition for a few reasons. 1. The technology for monorails travelling at 250 km/h currently did not exist. 2. There were no monorail systems anywhere in the world that that were that big (225 km in length between the two cities) 3. VIA Rail was proposing a high frequency slow train alternative to travel on existing tracks that was cheaper than high speed rail.

In 2012, a supportive organization "Solidarity Cooperative of the High-Speed Monorail" (Coop MGV) was formed in 2013 and the organization obtained its incorporation status as a coop.

Premier Philippe Couillard backed down from supporting the monorail idea and lost the next election.

Though a confidential Transport Canada document was leaked in July 2019 outlining the financial shortcomings of the Montreal-Quebec City VIA Rail high frequency rail project, the high-speed rail monorail project seemed unable to capitalize on this news.

Now, in 2020, the TrensQuébec website has not been updated since December 2017 and very little media of any kind is coming from the organization. The Coop MGV website and social media is current but it seems the idea of a high-speed monorail has 'hit a wall.'

2010 Parc Jean-Drapeau Monorail Proposal

In 2010 The paramunicipal agency that runs Parc Jean-Drapeau had a plan to build a monorail that would zip passengers to and from Île Sainte-Hélène and Île Notre-Dame as well from one part of the park to another. That plan was sitting on a shelf.

2018 SLC APM Group L'Animateur Monorail Proposal

On September 26th, 2018 Le Journal de Montreal reported that the SLC APM group, made up of the large accounting firm PwC and the engineering firm Arup, Loctrans and Solutrans, was proposing to build a 11.8 km monorail. It would be called " L'Animateur."

Group L'Animateur Monorail Proposal 2018. credit SLC-APM Group

It would have 14 stations between the convention center (in the city center) and La Ronde (on the island Sainte-Hélène). Station stops would include, among others, the casino and the new CHUM hospital, and would be connected to the Montreal metro as well as to the future Réseau express métropolitain (REM) line.

Estimated total cost: $1.1 billion. The infrastructure would be built in public-private partnership. This proposal is still in progress. According to the article the monorail would require the authorizations of many entities, including the City, the governments of Quebec and Canada, Six Flags, owner of La Ronde, as well as Loto-Quebec and the Caisse de depot.

MONTREAL QC EXPO MINIRAIL - 1967-PRESENT

Expo 1967. credit National Archives of Canada e011194216-v8

NOTE: The majority content of this section is taken from the 1967 publication titled, "The Minirail at Expo 1967 ("Expo 67") and Man and His World" - by the late Anthony Clegg with information courtesy of Mr. T. Zsenaty & Mr. D. Munro.

EXPO 67 is, by almost all standards, an unqualified success. Judged solely by the published attendance figures, this is unquestionably so, and the "Montreal World's Fair", the first Class "A" Exposition held under the auspices of the International Bureau of Expositions in the

Western Hemisphere, is targeting 34% more visitors than was originally estimated.

This tremendous acceptance of Expo by the people of Montreal and visitors alike is nonetheless the cause of one of the chief complaints - the long line-ups to gain admittance to the various pavilions and attractions. And nowhere has the waiting been longer and the queues more extensive than the line-ups to ride the MINIRAIL, open-air scenic rides around and above the Exposition Grounds.

MINIRAIL has captured the hearts of practically all its riders to an extent not even vaguely foreseen by the Expo planners. It was designed originally not as an amusement ride, but as a means of transit -- not Rapid Transit, but a secondary local transportation service to accommodate Expo visitors. The main route, colloquially known as "The Blue Minirail" from the colour of its cars, was laid out to pass close to almost all of the important points on the central part of Ile Ste. Helene and Ile Notre Dame.

The western part of Ile Ste. Helene and the La Ronde Amusement Park are served by similar but somewhat smaller MINIRAIL trains operating over independent loops and generally referred to as "The Yellow Minirail" due to the yellow canopies fitted over the diminutive open cars.

THE BLUE MINIRAIL

Screenshot of Blue Line. credit Bibliothèque et Archives nationales du Québec

On the major MINIRAIL circuit, there are six passenger stops strategically located close to the larger pavilions. The little trains pass beside the giant USSR pavilion, in front of the British and French structures, and thread their way among the large rock slabs at the Ontario exhibit. They pass directly through the centre of the pavilion of the United States of America, which is shaped like a huge crystal ball, although Minirail trains do not halt at this point and the closest passenger stop is some 1500 feet away. At one point, it passes between the exhibit of the Province of Ontario and that of the Province of Quebec,

Minirail Passengers pass under a series of scenic waterfalls, which cascade down from the main pavilion level to Regatta Lake and the St. Lawrence River below. At two points the line passes over itself, while at two other locations, the system ducks under the standard-gauge elevated rapid transit system "EXPO EXPRESS". Those responsible for laying out the MINI RAIL lines can take full marks for the picturesque way in which the diminutive trains wend their way around the scenic wonders of the exhibition park.

Operating over the 4.2 miles of track on this major MINIRAIL line, there are 32 trains of nine cars apiece. Each train is 125 feet in length and has a maximum capacity of 102 passengers. Generally speaking, the track is elevated above the crowded streets and walk-ways, but, as mentioned above, dips down occasionally almost to the level of the St. Lawrence,

All the trains are crewless and operate somewhat on the idea of an extended horizontal elevator. They are painted blue and white and remind one of the attractive tramcars in Lucerne, Berne, or Basel. Indeed, they should, for they were designed by the Swiss firm of Maschinenfabrik Habegger in Thun, Switzerland, which company also supplied the running gear and the automatic control equipment. Carbodies and superstructure were fabricated in Canada by the Truscon Division of Hawker Siddeley, in Montreal.

Building the monorail at Expo 1967
- Library and Archives Canada e011178101

Trackage was constructed on the Expo site by Dominion Bridge Company to Habegger specifications, with Herter Todd and Myer as local consultants. It consists of double "I" beams, 21 ¾" apart, supported on steel A-frames at 50-foot centres on straight sections and at 34-foot centres on curved sections. Track height ranges from ground level to a maximum of 40 feet above the surroundings. The ruling grade is 10%, and the minimum track radius is fifty feet.

Bogies between the passenger cars ride on two 27" pneumatic tired traction wheels and are guided by four horizontal stabilizing wheels of 9 ½ " diameter. Other supplementary guide wheels and emergency rollers are provided. A differential gear drive transmits power to the traction wheels from a 7 ½ H.P. D.C. electric motor, giving an average operating speed of about 7 ½ miles per hour. Maximum speed is in the vicinity of ten miles per hour.

Although the traction motors operate on D.C., current supply in the rails is at 440 volts A.C., 60 cycle, 3 phase. This is converted to Direct Current by a motor-generator set in the leading unit of each train. Leading units also contain the automatic train control mechanism, which governs all the trains' operations except station starting. The only manual controls on the system are the attendant-controlled buttons which permit trains to depart from the stations.

The four stations on the circuit are named after the nearest major pavilion or well-known landmark for which a bi-lingual designation can be employed: i.e., Metro; Agriculture; Theme; Canada. The fact that the trains stop at Agriculture and Theme both on the Metro-to-Canada leg of their run, and also on their return trip, makes six stops for a full circuit.

THE YELLOW MINIRAIL

Basically, similar to the "Blue Minirail" system described in the preceding paragraphs, the "Yellow Minirail" is divided into two loops -- one of 1.1 miles length serving the western end of Ile Ste. Helene - the other 1.3 miles long circling the Amusement Park area of La Ronde. There are twelve trains on each of these two lines, with sixteen cars per train.

In spite of the greater number of passenger-carrying units per train, the total length of the "Yellow" trains was only 105 feet, compared to the 125 feet-long trains on the "Blue" system. Comparative passenger capacities are 60 for the "Yellow" trains and 102 for the "Blue".

Track structure is slightly different in detail, but basically similar on all lines. Mojan Ltee were the Montreal contractors for the erection of the small minirail's trackage.

Historically, the European-built "Yellow" trains have a more chequered career than their Montreal-built "Blue" cousins. They were constructed in Switzerland by the Habegger firm at Thun in 1964 for the Swiss National Exhibition held that year at Lausanne. At that time, they were manually controlled but were converted to automatic operation for service at Expo 67.

Expo 67 Yellow Line - credit Glen Fairweather

FARE STRUCTURE

Rides on the Ile St. Helene and La Ronde minirail loops were forty cents per passenger, each line. Tariffs on the Ile Notre Dame system -- the more extensive "Blue" line -- were revised, and cost fifty cents per half-circuit. After operating for a number of weeks on a basis of "50 cents admission -- ride where you wish" if you could get on, the authorities inaugurated a policy of forcing everyone off at the Metro and the Canada stations. This made riding the MINIRAIL considerably more expensive and somewhat less popular.

The queues to gain admittance to this popular attraction are of reasonable proportions and the stations are not so completely overwhelmed by intending passengers. (Personally, I feel, however, that it would have been preferable to allow a complete round trip for the advance payment of one dollar.)

The complete MINIRAIL system was owned outright by the Canadian Corporation for the 1967 World Exhibition. First intentions were to assign this secondary transportation facility to a concessionaire, but this plan was later changed and the MINIRAIL is operated as part of the Expo '67 services. Its popularity seems to have justified this change of programme, and it was reported at the end of July 1967 that over seven million passengers had travelled on the Minirail lines during the first three months of operation.

PHOTO -

POST EXPO 1967

After the end of Expo 67, a new exhibition at the site, "Man and His World," struggled for years. In 1976 part of the site was repurposed as the rowing/canoe basin for Montreal's Summer Olympics. By 1984, due to fires, vandalism and neglect little was left of the original Expo 67.

The Blue Line continued on to 1973 when it was closed.

The Yellow Line on Saint Helen's Island west end operated until the end of 1981.

The other separate Yellow Line on Saint Helen's Island at La Ronde Amusement Park still operates. Since 2001, the park has been leased to the New York amusement park company Six Flags.

Montreal QC Expo 1967. Blue Line Around Bend.
credit Bibliothèque et Archives nationales du Québec

GRANBY QUEBEC
MONORAIL 1980 - PRESENT

Granby Zoo Monorail 2008 Photo credit - Martin Ujlaki

The Granby Zoo, in Granby Quebec, was founded in 1953. It is located about 70km from Montreal

The Granby Zoo monorail was built in 1979 and opened in May 1980. It was bought from Pierre Goulem of Amusements Goulem Inc.

To ride on the monorail around part of the zoo there is a small additional cost as it is not included in the Zoo's general admission.

You could in one day ride the La Ronde monorail in Montreal and then go to Granby Zoo and ride their monorail. Then you have been on both monorails still operating in Canada!

Granby Zoo Monorail 2006 credit - Sébastien Savard

CAVENDISH PEI - RAINBOW VALLEY MONORAIL - 1996 - 2005

Rainbow Valley monorail. credit Earle Davison

Earle Davison's Rainbow Valley Amusement Park was a fun thing to do when visiting Cavendish in Prince Edward Island. Most of the rides and animation figures were designed and built by Earle and staff at the park.

The park opened in 1969 and operated for 36 years until it closed in 2005.

In 1995-96 they built the mini monorail for the park. Looking back Earle stated, *"I realize that this was quite a project for our shop to tackle. I was the dreamer, the designer and the creator of this ride."* (Phone conversation with Earle)

Building the monorail track. credit Earle Davison

Engineering diagrams were done in Charlottetown for $150 by an engineer. Visiting Montreal, Earle found a company to make the iron for the rails and another Montreal company that could bend the iron to specifications.

Long-time collaborator, and excellent artist, Mary Smith was the person that designed the cars. The cars were metal frames with Fiberglass bodies built at Rainbow Valley.

Debra and Dylan Langan at Rainboaw Valley. credit author's collection

The cars were powered by a small air cooled Honda Motor, driving a small hydraulic Pump, the power went through a foot pedal valve and then to a small hydraulic Motor on the drive wheel. It is a U-drive monorail that required one of the riders in the whale-themed cars to press down on an accelerator to move along the track.

There was a valve that the riders did not see but allowed the cars to reverse if needed. On the outside of the car there was another valve, and the operator of the ride would turn it when the car came into the gate and that stopped the car. When the car was loaded the operator would turn this valve then the rider could step on the foot valve and be on their way.

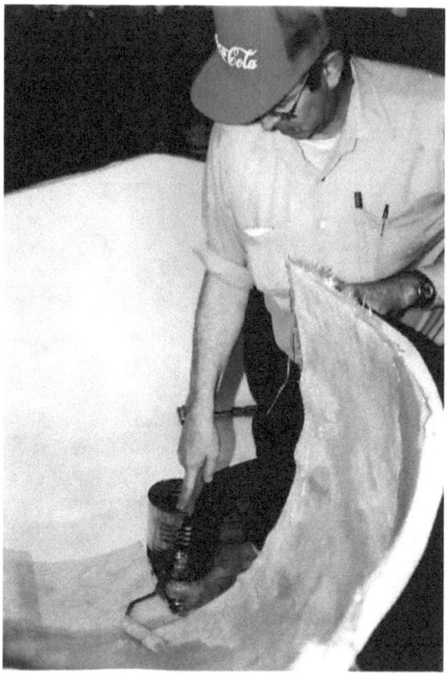

Earle working on car. credit Earle Davison

Reminiscing Earle stated. *"I was very lucky to have Mary on my team, she was an excellent artist and could draw things to scale, all I had to do was tell her what I was seeing in my mind and she could draw it for me. Once I had a picture of something, I could build it. With no engineering training, I drew out a lot of my creations on the back of a 12 month Caterpillar Equipment calendar. Mary would create a scale drawing that I could work from.*

Everything in that park except for the iron rides were built in our shop, and as I look back it was mostly a labour of love. I loved to build things and the park gave me a place to use that talent. Now in my 84th year I still build things for my son's projects."

He added, "I loved working with fiberglass, you could do so much with it and when you were in the park you must have seen that."

I visited the amusement park in 2003 with my family. We rode the monorail and took some photos. Little did I know that in a couple years the park would be closed.

Kate Langan at entrance to monorail. credit author's collection

When speaking with Earle about the monorail, he revealed an amazing fact. The complete monorail system exists and is still being kept in storage!

Maybe one day the Rainbow Valley monorail will exist again.

DARTMOUTH, NS - OTHO A. GIBBONS
AWARD WINNING MONORAIL PROPOSAL 1967

The Toronto Daily Star on March 3rd, 1967 had an article on Dartmouth's Otho A. Gibbons winning an international award of merit for his proposed solution to traffic congestion and highway fatalities.

Gibbons' idea was a mobile storage transit system by which automobiles would be carried inside giant gondolas traveling at high speed on an elevated monorail.

Gibbons worked on details of his plan for years before applying for his patent in 1959.

His car-carrying monorail won the award at the second International Inventors and New Products Exhibition held in New York City.

REFERENCES

British Columbia Wenner-Gren Monorail Proposal - 1957

(1957-03-09) Top engineers back monorail. *The Province*

(1957-02-13) Big plan foreseen to develop north. The Vancouver Sun

Ryan, B. (1957-02-13) Canadian investors will be welcomed. The Province

O'Leary D. (1957-02-13)180 MPH monorail, pulp mills planned. The Vancouver News-Herald

(1957-02-15) New legislation needed to file monorail plan. The Province

Province Ottawa Bureau (1957-02-15) Monorail could force subsidy. The Province

Province Victoria Bureau (1957-02-19) Monorail cars can be interchanged. The Province

(1957-02-19) Track could be changed. The Vancouver Sun

Fletcher B. (1957-02-21) It's your business. The Vancouver Sun

Fletcher B. (1957-03-11) Economics decide type of Wenner-Gren Railway. The Vancouver Sun

Metcalfe B. (1957-08-16) Survey progress pleases directors. The Province

(1957-10-14) Flooding to shift monorail line. The Vancouver Sun

Ryan, B. (1957-10-22) Trade mission impressed by ride on monorail train. The Province

(1959-09-07) Monorail plans reported off. Winnipeg Tribune

Vancouver - Wells Coates Proposal 1957

Cantasuzino, S. (1978) Wells Coates: A Monograph, Gordon Fraser, England

Cohn L. (1999) The door to a secret room - A portrait of Wells Coates. Scolar Press, England

Obituary (1959-06-19) Dr. Wells Coates, noted architect. Democrat and Chronicle, Rochester

Vancouver Bilger Monorail Proposal 1972

Bilger A.S. (1947-08-18) US2488523A Patent Anson S Bilger monorail wheel. US Patent Office

Bilger Monorail International Inc (PAM 1972-62) Bilger monorail preliminary proposal. City of Vancouver Archives

(1970-10-20) Before council today, monorail offer. The Province

(1972-04-06) Monorail backers seek new pitch. The Vancouver Sun

Finlay M. (1972-09-07) Monorail system shredded by planners. The Vancouver Sun

(1972-09-08) Morrison backs monorail pitch. Richmond Review

Vancouver Expo Monorail - 1986

Alton Towers Amusement Park History Source - **https://towersstreet.com/construction/monorail/**

Rice D.G. (Winter 1988) The Future of Transit Technology: Lessons From EXPO 86. Journal of Advanced Transportation, Vol. 21

Chan C. (2016-05-14) Expo 86: When Vancouver wooed the world — 30 photos, 30 years later. The Vancouver Sun

Calgary Brennan Monorail Exhibition - 1911

The Brennan Gyro-Monorail - http://www.douglas-self.com/MUSEUM/LOCOLOCO/brennan/brennan.htm

(1911-06-30) Monor-rail cars for state fair. Des Moines Register

Zakrison A. (2019-08-11) Twitter post on 1911 Brennan Monorail

Calgary (Ogden) Monorail 1911-1917

Peach J. (1982-07-10) Monorail idea stuck in the station since 1911. Calgary Herald

Zakrison A. (2019-08-11) Twitter posts on 1911-17 Alberta Monorail company- no sources in tweet for various old newspaper articles posted.

Calgary TranSyt Monorail Proposal 1982

Schuler D. (1982-05-28) City offered monorail deal. Calgary Herald

Morton P. (1982-10-22) Monorail test may start in spring. Calgary Herald

Martin D. (1983-03-21) City's monorail may be rerouted. Calgary Herald

Motherwell C. (1983-03-23) Council endorses monorail. Calgary Herald

Ross E. (1983-03-31) Monorail test talks progress. Calgary Herald

Morton P. (1983-04-14) New monorail test site sought. Calgary Herald

Motherwell C. (1983-12-02) Monorail squabble renewed. Calgary Herald

(1983-12-03) Monorail deal for Calgary denied but project not ruled out yet. Edmonton Journal

Priegert P. (1984-04-22) City loses test track. Calgary Herald

Richardson L. (1993-03-30) US5197390 Patent Suspended Vehicle Orientation Adjustment. US Patent Office

Calgary Other Monorail Proposals

(1951-11-01) Franchise is granted by Monorail Corporation. Daily Independent Journal California.

Cooper G. (1963-03-01) Calgary-Banff Latest Olympic. Edmonton Journal

Simon & Curtin (December 1967) Transit for Calgary's Future. City of Calgary Archives

Edmonton - Houston Monorail Corporation Proposal - 1962

(1962-02-14) Texas firm prepares a monorail plan for city. Edmonton Journal

(1962-02-22) Monorail firm plans city visit. Edmonton Journal

(1962-03-02) Monorail said sleeping giant. Edmonton Journal

(1964-05-29) City may get monorails. Edmonton Journal

Edmonton - Regina - Winnipeg - Royal American Midway - 1966

(1966-07-20) Two rides on opposite scale. Edmonton Journal

(1966-07-28) Monorail new ride. The Regina Leader-Post

Edmonton Alberta Monorail Ltd Fort Saskatchewan Proposal - 1980

Lambright, B. (1982-11-19) Bypassing car wars a $270,000 dream. Edmonton Journal

(1983-05-27) Monorail dream riding high. Edmonton Journal

Winnipeg Monorail Proposal 1962

Hills N. (1962-01-11) Transit chief has eye on portage monorail. Winnipeg Tribune

Winnipeg Stephen Juba Monorail Proposal 1972 - 1977

MacKay D. (1972-03-23) Monorail transit for city studied. Winnipeg Free Press

Editorial (1972-03-27) Pipe dream. Winnipeg Free Press

MacKay D. (1972-05-09) Newspapers spoiling monorail plan. Winnipeg Free Press

(1972-05-17) Monorail plan faces crucial vote. Winnipeg Free Press

(1972-05-18) Firm describes monorail. Winnipeg Free Press

Lisoway B. (1972-05-27) Secrecy at city hall. Winnipeg Free Press

Campbell R. (1973-02-20) $4 million a mile transit. Winnipeg Free Press

(1973-10-23) Juba's plan hit. Winnipeg Free Press

(1973-10-27) Schreyer says millions available for urban transit. Winnipeg Free Press

(1974-01-02) Monorail. Winnipeg Free Press

Wielaard R. (1975-06-27) Agrees to revive monorail scheme. Winnipeg Free Press

(1975-12-03) Monorail, light train or bus lane are favoured. Winnipeg Free Press

(1976-08-06) Juba has a new vision, luxury cars. Winnipeg Free Press

(1976-12-23) Juba raps proposal for corridor diesels. Winnipeg Free Press

(1977-01-07) City eyes electric trolleys. Winnipeg Free Press

(1977-04-07) Council revives monorail idea. Winnipeg Free Press

Pona S. (1977-08-12) Council nears agreement on monorail. Winnipeg Free Press

(1977-10-08) Juba wins verbal tussle on monorail. Winnipeg Free Press

Editorial (1977-10-27) Monorail gone with the wind. Winnipeg Free Press

Ontario Southern Railway Monorail - Ridgeway ON - 1896 - 1898

(1900) 200 miles per hour, Holly Standard NY

Beecher L. (1902-03-11) US695137 Patent Monorailway. US Patent Office

Wayne J. Days When You Could Ride the Monorail to Crystal Beach. Author Collection

Opinion (2011-10-06) Did you know... Fort Erie once had a monorail? Fort Erie Post

Niagara Falls Monorail Proposal 1963-2009

(1963-12-12) Monorail planned. Niagara Falls Evening Review

(1964-01-14) Monorail project hits snag, won't be read for 64. Niagara Falls Evening Review

(1965-02-17) Council Clears Niagara Falls Monorail Plan. The Montreal Gazette

(1965-02-17) 10-block monorail franchise is approved by Niagara Falls. The New York Times

(1965-02-17) Monorail for Niagara Falls. Edmonton Journal

(1968-07-23) Proposed monorail plans shown. Niagara Falls Review

Hvilivitzky J. (1969-07-18) Monorail to start at Queen for tourists and communities. Niagara Falls Review

O'Neill C. (1969-12-02) No legislation to permit monorail construction. Niagara Falls Review

(1970-01-07) Niagara Monorail ok'd. Daily News NY

Cullen J. (1970-02-15) Niagara Falls firm is hired to engineer area monorail. Niagara Falls Review

Gannon B. (1970-02-24) Monorail bill expected soon. Niagara Falls Review

Cullen J. Cullen J. (1970-02-15) Niagara Falls firm is hired to engineer area monorail. Niagara Falls Review

(1970-07-09) Work on monorail system won't resume until mid-July. Niagara Falls Review

Hvilivitzky J. (1971-01-09) Monorail may be abandoned. Niagara Falls Review

(1996-03-26) $140 million Falls monorail idea makes stop at region. Niagara Falls Review

(1997-06-28) People mover moving along. The Tribune

Kilpatrick, K. (2000-02-16) Monorail system to connect Niagara tourist attractions. Globe and Mail

Larocque C. (2005-06-23) Cost prohibitive for elevated people mover. Niagara Falls Review

(2009-04-24) City bids farewell to monorail, people mover to come in the form of buses. Niagara This Week

Toronto - 1954- 2014

(1955-09-21) Monorail for Bloor St. Toronto Daily Star

(1956-03-27) No monorail for Toronto, Lamport. Toronto Daily Star

(1958-04-25) Lamport explains why monorail not answer. Toronto Daily Star

(1958-05-01) Don't make Toronto a monorail guinea pig. Toronto Daily Star

(1958-05-03) So much for being a fighter. Toronto Telegram

(1958-05-03) The monorail flop. Toronto Daily Star

(1958-05-07) Pro-monorail members surprised by support. Toronto Daily Star

(1958-05-08) Monorail passengers claim system works. Weston Times and Guide

(1958-05-21) Council sees Toronto to Hamilton monorail. Toronto Daily Star

(1958-06-09) Monorail schemes rapped by Duncan Toronto Daily Star

(1958-06-18) Vote subway, dump monorail plan. Toronto Daily Star

(1958-06-21) No monorail for Toronto. Toronto Daily Star

(1958-06-30) Avro tells city monorail not good idea. Toronto Daily Star

(1958-08-06) Monorail seen to Malton, New Woodbine, Rexdale. Toronto Daily Star

(1959-05-26) Four reeves can't win monorail support. Toronto Daily Star

Nicol E. (1959-07-13) Is monorail way out of the nightmare of big city's traffic snarls? Toronto Daily Star

(1961-05-25) Monorail system urged for metro. Toronto Daily Star

(1963-09-27) Wants study of monorail for metro. Toronto Daily Star

(1964-06-03) Monorail suggested. Toronto Daily Star

(1965-01-25) Sky-ride monorail talked for 1967 Ex.. Toronto Daily Star

(1965-08-16) Too expensive and too rough Ex monorail idea scrapped. Toronto Daily Star

(1966-07-20) Monorail costs skyrocket. Toronto Daily Star

Wells D. (1969) A Million-Dollar Look for the CNE. Toronto Public Library

(1970-02-25) 100 m.p.h. city transit predicted. Toronto Daily Star

(1971-06-10) Air-cushioned commuter train studied by Ontario experts. Toronto Daily Star

(1971-11-30) After three years the attack sharpens. Toronto Daily Star

(1972-03-03) Hovercraft, monorail may whisk passengers to new airport. Toronto Daily Star

(1973-11-27) TTC: 2 rails better than GO-Urban. Toronto Daily Star

(1973-11-27) Largely ignored TTC report raises doubts about monorail. Toronto Daily Star

(1973-12-11) Monorail not harmful council told. Toronto Daily Star

(1975-01-02) Major transit changes predicted for Metro. Toronto Daily Star

(1982-03-02) Mississauga designs its future. Toronto Daily Star

(1982-04-20) Speeding bullet transit on exhibition. Toronto Daily Star

(1982-04-30) North York firm building high-speed monorail train. Toronto Daily Star

(1996-11-06) Theme park in running for CNE grandstand. Toronto Daily Star

(2011-08-30) City wants to seize control of port lands National Post

(2011-08-31) There is fog in Doug Ford's vision. Toronto Daily Star

Editorial (2011-09-01) Ride the monorail to waterfront ruin. Toronto Daily Star

Kalinowski T. (2011-09-05) The vision of a one-track mind. Toronto Daily Star

Rider D. (2011-09-10) Port Agency fluid on plans. Toronto Daily Star

Rider D. (2011-09-15) Ford ally joins experts in rejecting waterfront plan Toronto Daily Star

James R. (2011-09-22) Lessons learned from Ford's waterfront rebuff. Toronto Daily Star

Doolittle R. (2011-09-24) Has Mayor Ford lost his grip. Toronto Daily Star

(2014-09-16) A snapshot of the new candidate. Toronto Daily Star

Montreal Monorail Proposals 1947-2020

(1947-11-07) Adoption is proposed of overhead monorail. The Gazette

(1953-08-22) Montreal to study monorail system for city. The Gazette

(1953-08-29) Montreal plans in US to be studied by city. The Gazette

Peters C.H.(1953-09-19) Monorail nice - but doesn't solve traffic jams. The Gazette

(1954-03-23) Subway here is enough people ask. The Gazette

Editorial (1955-09-16) This seems a little far-fetched. The Gazette

(1955-09-16) Campeau urges monorail survey. The Gazette

(1961-01-24) Top priority for subway. The Gazette

(1962-12-18) Extension of subway to airport? The Gazette

(1966-05-11) Eastern monorail proposed. Edmonton Journal

Radwanski G. (1966-05-11) Monorail-Utica monorail planned by N.Y syndicate. The Gazette

Radwanski G. (1966-05-11) Monorail-Utica line given boost. The Gazette

Palmer A. (1966-11-22) Mono Talk. The Gazette

(1966-12-08) Utica monorail boosters in town. The Gazette

(1967-06-13) Engineers, Economists study proposed N.Y. - Montreal link. The Gazette

(1968-08-28) Monorail, metro extension urged. The Gazette

(1968-08-30) Anjou monorail line possibility studied. The Gazette

Windeyer K. (1968-08-30) Airport monorail studied. The Gazette

Miskew B. (1970-10-30) High-speed monorail proposed for airport. The Regina Leader-Post

Laurent R. (1974-04-11) South shore study of transport ordered. The Gazette

Leckner R. (1987-03-16) A solution to bridge congestion? The Gazette

Scott S. (1989-01-25) Cote laughs off fears of more urban sprawl. The Gazette

(1989-01-27) A bold idea for Laval. The Gazette

King M. (1989-08-28) Bourassa promises a study of transit link to Laval. The Gazette

Lamey M. (1992-12-01) CN proposes monorail links to South Shore, Laval. The Gazette

Sheperd H. (1992-12-03) Backers of ice-bridge route turn up the heat. The Gazette

Sheperd H. (1992-12-10) Ice-bridge backers try to appease opponents. The Gazette

Lamey M. (1993-04-20) CP Rail proposes joint off-island commuter-train network, Laval. The Gazette

Lamey M. (1993-04-20) Joint CN-CP train plan has $100-million price tag. The Gazette

Sheperd H. (1994-09-08) Group proposed monorail across river. The Gazette

Sheperd H. (1995-10-05) Buses on ice bridge best plan study says. The Gazette

Phaneuf I. (2000-01-19) Minister plays it cool on proposed bridges. The Gazette

Gamble D. (2000-01-30) Traffic solution outlined. The Gazette

Gyulai L. (2000-02-11) Harel likes light-rail plans. The Gazette

Leong R. (2003-10-13) New casino? No dice, Sequin. The Gazette

Dobrota A. (2005-06-23) Casino - Cirque du Soleil enlisted as creative partner for ambitious plan. The Gazette

(2006-04-15) Exhibition hall dead, stadium on hold. The Gazette

(2010-01-11) Monorail anyone? The Gazette

(2010-10-09) Train link to airport: proposals and promises go way back. The Gazette

Dutil, Y. (2013-04-08) The monorail against the Quebec inferiority complex. https://www.sciencepresse.qc.ca/blogue/2013/04/08/monorail-contre-complexe-dinferiorite-quebecois

Riga A. (2013-04-11) New federal money for light-rail on Champlain. The Gazette

D'Alimonte M. (2017-11-27) Quebec Wants to Build A Monorail from Montreal To Quebec City, https://www.mtlblog.com/en-ca/news/quebec-wants-to-build-a-monorail-from-montreal-to-quebec-city

(2018-01-15) The too daring bet of a monorail between Montreal and Quebec? Le Devoir

Fortin J.L. (2018-09-26) Another monorail is proposed for Montreal. Le Journal de Montreal

(2019-08-13) Monorail to Laval: How Montrealers blocked a plan to link Laval to the Montreal metro system. http://coolopolis.blogspot.com/2019/08/monorail-to-laval-how-montrealers.html

TrensQuebec Website - http://www.trensquebec.qc.ca/

Montreal QC Expo MiniRail - 1967-Present

(1963-03-29) Dazzling future viewed for St. Helen's fair site. The Gazette

(1964-05-22) Firm seeks monorail sale to world fair. The Gazette

(1964-06-04) $4 million monorail discussed with Expo. The Gazette

(1964-08-29) Expo transit to be electric, elevated. The Gazette

(1965-02-24) Consortium enters successful bid for monorail transport at Expo 67. The Gazette

The Minirail at Expo 67 and Man and his World by Anthongy Clegg. From information courtesy of Mr. T. Zsenaty & Mr. D. Munro.

https://pdfslide.net/documents/the-minirail-at-expo-67-and-man-and-his-world-by-anthony-clegg.html

Habegger Swiss Company involved with 1967 montreal. monorailhttps://amusementparkives.com/2017/02/14/maschinenfabrik-habegger/ .

Granby Quebec Monorail 1980- Present

Email correspondence with Zoo staff by author

Cavendish PEI - Rainbow Valley Monorail - 1996-2005

Phone and email communication with founder of Rainbox Valley monorail Earl Davidson

Dartmouth NS - 0tho A. Gibbons Award Winning Monorail Proposal 1967

(1967-03-03) All aboard the train-and bring your car. Toronto Daily Star

Expo 86 Vancouver Track Removal. credit Reinhold 2 Kocaurek.com

Made in United States
Troutdale, OR
11/03/2026

56844689R00076